MATH

FOR THE COSMETOLOGY AND ESTHETICS PROFESSIONAL

DOUGLAS STULTZ

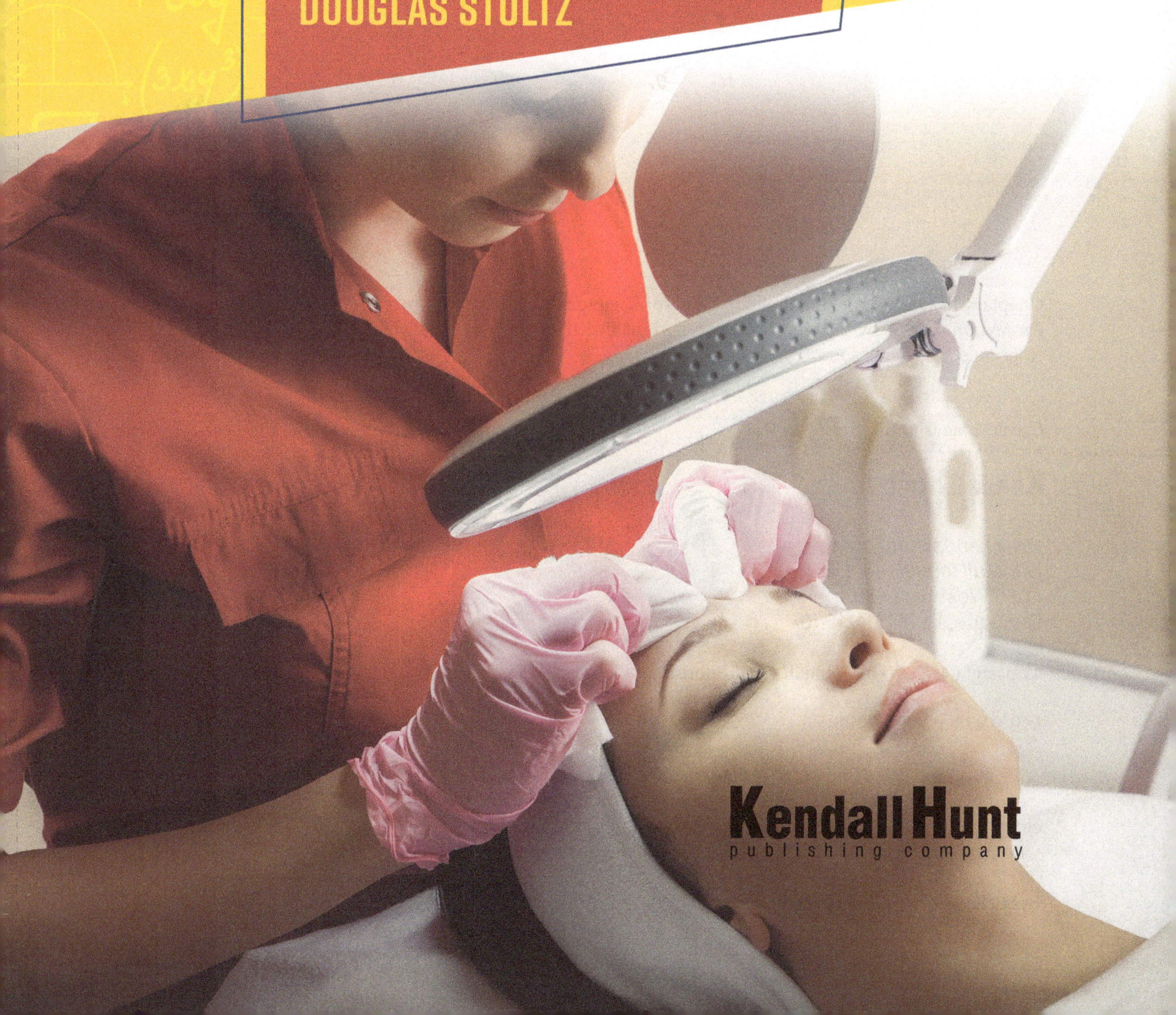

Kendall Hunt
publishing company

Chapters 1, 2, and 3 and the Practice Tests for Chapters 1, 2, and 3 are from *Math for Industrial Professionals, 2/E* by Douglas Stultz. Copyright © 2019 by Kendall Hunt Publishing Company. Reprinted by permission.

Cover images © Shutterstock, Inc.

Kendall Hunt
publishing company

www.kendallhunt.com
Send all inquiries to:
4050 Westmark Drive
Dubuque, IA 52004-1840

ISBN 978-1-7924-2305-5

Published in the United States of America

Contents

Foreword

As with any profession, mathematics plays a part in the cosmetology or esthetics profession. During a typical day in one's business, the professional is faced with a myriad of math related questions and problems that need to be solved. From taking a ½ inch parting from a haircut, to converting from milliliters to fluid ounces in products volume, to scheduling hours of appointments into a day schedule, to figuring the markup on a makeup kit for sale, to planning for a compound loan or an annuity investment, to calculating taxable income and federal taxes for payroll. YOU ARE DOING MATH!

The text, Mathematics for Cosmetology and Esthetics Professional, is designed to help students solidify a mathematics foundation as they face their cosmetology or esthetics professional program and career. The student must have confidence in the use of the basic mathematic and algebraic principles and properties as well as an understanding on how he or she can apply them in creating and solving algebraic expressions and equations and in using formulas and measurements. Each chapter provided instruction, demonstrations, practice work, and assignments for additional practice.

It is extremely important students have confidence in mathematics basics, such as fractions, decimals, order of operations, and raising to a power, as they are used frequently in of problem-solving and calculations involving measurements and business operation. Chapter 1 provides instruction on mathematical operations (adding, subtraction, multiplication, division, and raising to a power.) In addition, there is instruction and practice working with fractions and decimals as well as proper order of operations.

As the mathematics used in problem solving in the cosmetology or esthetics profession often require using variable formulas, a basic understanding of algebraic concepts is needed. Chapter 2 provides instruction on creating and simplifying algebraic expressions. For when these algebraic expressions are used in equations, the chapter demonstrates how to use mathematic properties of equality to solve for the variable in the equation. Also covered are ratio and proportions, rational variable equations, calculating percent's, evaluating and solving formulas.

It is critical that students in cosmetology or esthetics programs understand and be able to use standard units of measure. Chapter 3 covers systems of measurement, primarily focusing on the Customary System utilized primarily in the United States and the Metric System favored by the rest of the world. Measurements covered will include length, volume, and mass/weight. Description of the various units of measure will be demonstrated and instruction will be given on conversions within measuring families and between different measuring families. Included as well is a section on solving mixture problems.

When running a cosmetology/esthetics business, the professional needs to be able schedule appointment of customers, meetings, and other time based activities into the business day. The student needs to be able to calculate hours and minutes and plan accordingly. Chapter 4 introduces different time systems and presents ways to add and subtract hours and minutes in a practical time management process.

When running a business, the cosmetology/esthetic professional offers services and items for sale to customers and are faced with costs and expenses. Given those sales and costs, the professional needs to be able to show a growing profit in order to continue and thrive in the business. Chapter 5 provides the student instruction on how to apply sales tax, markups, and markdowns on products and services for sale. The chapter also covers how to calculate sales and costs to determine profit, and ways to represent this data. The chapter will provide a variety of business scenario examples and practice problems to help the student gain confidence in using the business related math.

In one's cosmetology/esthetics career and one's personal life, the professional is often faced with a need to borrow money to help his or her business, and may want to invest money for a future goal, such as a business improvement, down payment on a retail location, or to plan for the future and retirement. Chapter 6 covers interest bearing loans and investments. The chapter provides instruction on using formulas to calculate simple interest, compounding interest, and annuities. The student will gain confidence through examples and practice problems so they can handle the simple basics of business and personal financial planning.

As a business owner, the cosmetology/esthetics professional is often responsible for doing payroll for his or herself and for employees. Chapter 7 is designed to provide the student instruction on figuring gross income, taxable income, Federal Income Tax, payroll deductions and net income, as well as an employer's federal tax obligations. The chapter will provide a variety of compensation scenario examples and practice problems to help the student gain confidence in using completing a business payroll.

To assist in the learning, there are practice tests with an answer key at the back of the book.

It is hoped that through this review and instruction will assist the student in their course of study by giving them tools to use and confidence they need to be successful in their cosmetology or esthetics career.

CHAPTER 1

Order of Operations

1-1 FRACTIONS/DECIMALS

ALL FRACTIONS ARE CONSIDERED TO BE EXACT.

A fraction is a number that describes a part of a whole. It is also the relationship between two values by division. The fraction appears in this way: one of the values, called the numerator, is placed over the other value, called the denominator, with the fraction line between them acting like a division sign.

For example $\frac{2}{3}$, with 2 the numerator and 3 the denominator, is the relationship of 2 to 3 by division, or represents 2 divided by 3.

When the fraction "number" itself is left undivided, we say the number is EXACT.

When the fraction "number" is then divided, resulting in a decimal value, we consider the resulting decimal at best an APPROXIMATE value. For example, $\frac{2}{3}$ is exact, but when 2 is divided by 3 we get .6666... or $.\overline{6}$, a decimal that repeats the number 6 indefinitely. To come up with a number value, we can, at best, round the number to some decimal place value. We can say $\frac{2}{3} = .6666\ldots \approx .67$ rounded to the nearest hundredth .... So the best we can say is the decimal value is approximate. Even when fractions result in whole numbers or decimals that terminate which appear "exact", such as $\frac{4}{2} = 2.0$ or $\frac{1}{2} = .50$, we still say the decimal is approximate because there is still a call for some determination of rounding to a value. As in, 2.0 rounded to the nearest tenth, 2.00 rounded to the nearest hundredth, and so on.

There are other operation forms that have exactness versus approximation reasoning, such as

- Roots, example $\sqrt{2}$ is exact; it's number value is 1.414... which is approximate
- logarithms, example log 50 is exact; its number value is 1.698.... which is approximate
- trig functions, example sin 50° is exact; .7660.... is approximate

WHEN TO USE FRACTIONS (or other operation) AND WHEN TO USE DECIMAL IN SIMPLIFYING EXPRESSIONS AND/OR SOLVING EQUATIONS.

The Rule of thumb says "ANSWER IN THE SAME FORM YOU ARE GIVEN!"

Example, Joe mixes ½ quarts of skim milk with 1/4 quarts of whole milk. How much new milk does he have?

½ + ¼ = ¾, so he would have ¾ quarts of new milk.

However, it also true that ½ + ¼ = .5 + .25 = .75. So it would correct to say he has .75 quarts of new milk, but it would be BETTER and proper to say 3/4 quarts instead.

1-1 Assignments

1. Which of the following are EXACT numbers and which are APPROXIMATE numbers
   a) .56 ____________
   b) $.17\overline{8}$ ____________
   c) $\frac{5}{6}$ ____________
   d) $\frac{21345}{14992}$ ____________
   e) .5000 ____________
   f) $\frac{1}{2}$ ____________
   g) $\sqrt{7}$ ____________
   h) $2\frac{3}{4}+5+.0004$ ____________
2. True or false
   a) Some decimal numbers can be considered exact. ____________
   b) All rational fractional numbers are considered exact. ____________
   c) Ideally, if a problem uses fractions with math operations (adding, subtracting,..), it doesn't matter if your answer solution is in a fraction or a decimal form ____________
   d) If a number is given as a fraction plus a decimal, then the number is not exact. __________
   e) Unless otherwise instructed, the sum $\frac{1}{2}+3\frac{1}{4}$ should be recorded as $3\frac{3}{4}$ and not 3.75 ________

1-2 ADDITION AND SUBTRACTION OF FRACTIONS

ONLY LIKE QUANTITIES CAN BE ADDED OR SUBTRACTED.

We can add (or subtract) fractions only when they have the SAME denominator. We say these fractions with common denominators are LIKE fractions.

If you are adding (or subtracting) fractions with the same denominator it is easy: you simply add (or subtract) the numerators and put that sum (or difference) over the common denominator.

Such as $\frac{2}{7}+\frac{4}{7}=\frac{2+4}{7}=\frac{6}{7}$

Or $\frac{9}{15}-\frac{3}{15}=\frac{9-3}{15}=\frac{6}{15}$

When you want to add (or subtract) fractions with different denominators, we have to create equivalent fractions with the same denominator so they are like fractions. We can do so by multiplying the fraction(s) by fractional "identities". A fractional identity is a fraction that contains the same value in the numerator and denominator, making the fraction equal to 1. This works because any number divided by itself is equal to 1.

Examples of fractional identities are $\frac{3}{3}, \frac{500}{500}, \frac{y}{y}, \frac{a+b}{a+b}$ and so on.

We can change our fractions in our sum (or difference) to those with common denominators.

Example add $\frac{3}{4}+\frac{4}{5}$. We have uncommon denominators. 4 and 5. But if we multiply 4 by 5 we get 20 and multiply 5 by 4 we get 20, so we can get a common denominator of 20 by multiplying the numerator and denominator of $\frac{3}{4}$ by 5 and multiplying the numerator and denominator of $\frac{4}{5}$ by 4.

Here is what it will look like:

$\frac{3}{4}+\frac{4}{5}=\frac{3}{4}\cdot\frac{5}{5}+\frac{4}{5}\cdot\frac{4}{4}=\frac{15}{20}+\frac{16}{20}$

Do you see the fraction identities, $\frac{5}{5}$ *and* $\frac{4}{4}$?

Now we have a common denominator of 20, and adding

$\frac{15}{20}+\frac{16}{20}=\frac{15+16}{20}=\frac{31}{20}$

The same thing can be done with variable fractions.

Example: we want to subtract $\frac{5}{x}-\frac{2}{3y}$.

Since x times 3y = $3xy$ and 3y times $x = 3xy$, we can get a common denominator of $3xy$ by multiplying $\frac{5}{x}$ by $\frac{3y}{3y}$ and $\frac{2}{3y}$ by $\frac{x}{x}$,

so $\frac{5}{x}-\frac{3}{3y}=\frac{5}{x}\cdot\frac{3y}{3y}-\frac{2}{3y}\cdot\frac{x}{x}=\frac{15y}{3xy}-\frac{2x}{3xy}=\frac{15y-2x}{3xy}$

Practice

$\frac{2}{9}+\frac{5}{9}=$ $\frac{3}{5}-\frac{1}{7}=$

$\frac{3}{y}+\frac{4}{y}=$ $\frac{2}{3y}-\frac{4}{5x}=$

Answers: $\frac{7}{9}$; $\frac{16}{35}$; $\frac{7}{y}$; $\frac{(10x-12y)}{15xy}$

REDUCING FRACTION

The fractional identity also can be used to reduce fractions.

When the numerator and denominator share a common integer factor, one can "divide" that integer in both numerator and denominator, or dividing by a fraction identity.

Example, reduce $\frac{6}{15}$

3 is a factor of both 6 and 15, so we can write $\frac{6}{15}=\frac{6\div 3}{15\div 3}=\frac{2}{5}$

When there are no other common integer factors, we say the fraction is reduced.

HERE ARE SOME DIVISIBILTY RULES TO HELP YOU

Number	Rule
2	If the number is even
3	If you can add all of the digits of a number and that number is divisible by three. Example: 123. Add 1 + 2 + 3 = 6. Because 6 is divisible by 3, 123 must be divisible by 3
4	The last 2 digits are divisible by 4, or it is divisible by 2 twice. 124 is divisible by 4 because 24 is divisible by 4.
5	Ends in a 0 or 5
6	It is divisible by 2 and 3. An even number that follows the divisibility rule for 3.
7	No good rule. Ignore it and use the nearest number method discussed in the mental math section.
8	Same rule as 4 but for the last 3 digits, or it is divisible by 2 three times. 1,240 is divisible by 8 because 240 is divisible by 8
9	Same rule as 3, but the sum of the digits must be divisible by 9
10	If it ends in a 0.

Practice

Reduce these fractions

$\frac{10}{25}$ ________ $\frac{8}{12}$ ________ $\frac{16}{64}$ ________ $\frac{27}{45}$ ________ $\frac{132}{1731}$ ________

Answers: $\frac{2}{5}$; $\frac{2}{3}$; $\frac{1}{4}$; $\frac{3}{5}$; $\frac{44}{577}$

CONVERTING MIXED NUMBERS TO IMPROPER FRACTIONS

A **mixed number** is a combination of a whole number and a fraction and represents the sum of the whole number and the fraction. For example, $2\frac{1}{2}$ is the sum $2+\frac{1}{2}$.

A fraction in which the numerator is smaller than the denominator, like $^1/3$ or $^2/5$ is called a **proper fraction.** A fraction in which the numerator is larger than or equal to the denominator, like $^5/2$, $^{17}/3$, or $^6/6$ is called an **improper fraction.** To put it another way, a positive fraction with a value less than 1 is a proper fraction. A fraction with a value greater than or equal to 1 is an improper fraction.

To turn a mixed number into an improper fraction, multiply the whole number by the denominator (the bottom part), and add the result to the numerator (the top part). Then this sum goes over the original denominator.

Examples $4\frac{2}{3}=\frac{4\cdot 3+2}{3}=\frac{14}{3}$; $\quad 10\frac{5}{7}=\frac{10\cdot 7+5}{7}=\frac{75}{7}$

CONVERTING AN IMPROPER FRACTION TO A MIXED NUMBER

To write an improper fraction as a mixed number, divide the numerator (top part) by the denominator (bottom part). The quotient is the whole number, and the remainder is the numerator.

Examples $\frac{14}{3} \to 3\overline{)\,14}$ with quotient 4; -12; remainder 2, so $\frac{14}{3}=4\frac{2}{3}$

ADDING(SUBTRACTING) MIXED NUMBERS

When you add (or subtract) mixed numbers, you can add the whole numbers then add the fractions using fraction rules then convert the result into a new mixed number.

For Example: $2\frac{3}{4}+5\frac{1}{7}$

We can write $2\frac{3}{4}+5\frac{1}{7}$ as $2+\frac{3}{4}+5+\frac{1}{7}$

We can rearrange the trems $2+\frac{3}{4}+5+\frac{1}{7}$ as $2+5+\frac{3}{4}+\frac{1}{7}$

Adding the whole numbers we get $2 + 5 = 7$

Adding the fractions $\frac{3}{4}+\frac{1}{7}=\frac{3\cdot 7}{4\cdot 7}+\frac{1\cdot 4}{7\cdot 4}=\frac{21}{28}+\frac{4}{28}=\frac{21+4}{28}=\frac{25}{28}$

So $2\frac{3}{4}+5\frac{1}{7}=7+\frac{25}{28}=7\frac{25}{28}$

What about subtracting mixed number $12\frac{4}{5}-9\frac{2}{3}$

We can write this as $12+\frac{4}{5}-(9+\frac{2}{3})=12+\frac{4}{5}-9-\frac{2}{3}$

We can rearrange the terms of $12+\frac{4}{5}-9-\frac{2}{3}$ as $12-9+\frac{4}{5}-\frac{2}{3}$

Now the whole number is $12-9=3$

and the fractions, $\frac{4}{5}-\frac{2}{3}=\frac{4\cdot 3}{5\cdot 3}-\frac{2\cdot 5}{3\cdot 5}=\frac{12}{5}-\frac{10}{15}=\frac{12-10}{15}=\frac{2}{15}$

So $12\frac{4}{5}-9\frac{2}{3}=3+\frac{2}{15}=3\frac{2}{15}$

Practice

$7\frac{3}{8}+4\frac{2}{7}$ ______ $15\frac{3}{4}-6\frac{1}{9}$ ______

Answers: $11\frac{37}{56}$; $9\frac{23}{36}$

ESTIMATING USING MIXED NUMBERS

When you add (or subtract) mixed numbers, you can use the whole number part to estimate your result.

Example $12\frac{3}{8}+15\frac{2}{7}$ can be estimated to $12 + 15 = 27$. Since the fraction parts can be anywhere from 0 to 2, So we say that this results would be about 27 or 28.

We also use estimating when we round numbers that are decimals. Becoming proficient in quickly estimating is important when we want to check our final answers, get a feel for what our answer should be around, etc.

1-2 Assignments

Reduce these fractions

1. $\frac{8}{16}=$ **2.** $\frac{14}{18}=$ **3.** $\frac{28}{35}=$ **4.** $\frac{145}{150}=$

5. $\frac{54}{108}=$ **6.** $\frac{51}{63}=$ **7.** $\frac{924}{1016}=$ **8.** $\frac{4545}{9999}=$

Add or subtract fractions, simplify final fractions

9. $\frac{1}{9}+\frac{5}{18}=$

10. $\frac{1}{9}+\frac{7}{20}=$

11. $\frac{3}{7}+\frac{1}{15}=$

12. $\frac{1}{12}+\frac{1}{9}=$

13. $\frac{7}{8}+\frac{6}{7}=$

14. $\frac{11}{20}+\frac{11}{40}=$

15. $\frac{2}{5}-\frac{1}{5}=$

16. $\frac{2}{5}-\frac{5}{16}=$

17. $\frac{3}{4}-\frac{2}{15}=$

18. $\frac{7}{8}-\frac{7}{16}=$

19. $\frac{1}{2}-\frac{3}{16}=$

20. $\frac{9}{20}-\frac{1}{7}=$

21. $\frac{1}{b}+\frac{5}{a}=$

22. $\frac{1}{xy}+\frac{7}{2w}=$

23. $\frac{3t}{7}+\frac{5t}{14}=$

24. $\frac{t}{p}-\frac{y}{x}=$

25. $\frac{ab}{cd}+\frac{ef}{gh}=$

26. $\frac{1}{f}+\frac{h}{4}=$

27. $2\frac{1}{7}+9\frac{11}{15}=$

28. $4\frac{14}{15}+1\frac{7}{12}=$

29. $7\frac{3}{10}+7\frac{1}{16}=$

30. $7\frac{1}{7}+1\frac{2}{9}=$

31. $5\frac{4}{9}+8\frac{3}{7}=$

32. $9\frac{8}{9}+6\frac{4}{7}=$

33. $8\frac{5}{7}-2\frac{2}{15}=$

34. $9\frac{8}{9}-6\frac{1}{2}=$

35. $10\frac{1}{12}-7\frac{3}{4}=$

36. $5\frac{2}{5}-1\frac{1}{2}=$

37. $8\frac{7}{20}-4\frac{1}{4}=$

38. $7\frac{5}{9}-6\frac{3}{4}=$

1-3 MULTIPLYING FRACTIONS

When multiplying fractions, the rule is

MULTIPLY NUMERATORS ACROSS, MULTIPLY DENOMINATORS ACROSS

Example: $\frac{2}{7}\cdot\frac{6}{11}=\frac{2\cdot 6}{7\cdot 11}=\frac{12}{77}$

MULTIPLYING MIXED NUMBERS

When multiply mixed numbers, convert the mixed numbers to improper fractions then use fraction multiplication rule. Then reconvert the resulting improper fraction to a mixed number.

Example: $5\frac{1}{6}\cdot 3\frac{2}{9}=\frac{5\cdot 6+1}{6}\cdot\frac{3\cdot 9+2}{9}=\frac{31}{6}\cdot\frac{29}{9}=\frac{31\cdot 29}{6\cdot 9}=\frac{899}{54}$

Converting this improper fraction to a mixed number

$$\begin{array}{r} 16 \\ 54\overline{)899} \\ -54 \\ \hline 359 \\ -324 \\ \hline 35 \end{array}$$

so $\frac{899}{54} = 16\frac{35}{54}$ so $5\frac{1}{6} \cdot 3\frac{2}{9} = 16\frac{35}{54}$

CANCELLING OUT FACTORS

When we multiply fractions, we can "reduce" the multiplication by cancelling our common factors in numerators and denominators across the fraction factors, making the calculation easier, as well as provide a simplified fraction result.

Example: $\frac{4}{11} \cdot \frac{33}{2}$ as the 4 in the first fraction numerator and the 2 in the second fraction denominator share a factor of 2 we can cancel out 2 from each. Also the 11 in the first fraction denominator and the 33 in the second fraction numerator share a factor of 11, we can cancel 11 from each. So we get

$$\frac{\overset{2}{\cancel{4}}}{\underset{1}{\cancel{11}}} \cdot \frac{\overset{3}{\cancel{33}}}{\underset{1}{\cancel{2}}} = \frac{2}{1} \cdot \frac{3}{1} = \frac{2 \cdot 3}{1 \cdot 1} = \frac{6}{1} = 6$$

Practice: Multiply

$\frac{3}{11} \cdot \frac{1}{2}$ ________, $\frac{7}{12} \cdot \frac{36}{21}$ ________, $2\frac{5}{6} \cdot 3\frac{4}{5}$ ________, $\frac{2}{3} \cdot 7\frac{6}{7}$ ________.

Answers: $\frac{3}{22}$; 1; $10\frac{23}{30}$; $5\frac{5}{21}$

DIVISION BY QUANTITIES

A DENOMINATE NUMBER is made up of a numerical value followed by a unit of measure, such as 12 ft., or 345 lb., or 789 yr.

We can divide denominate numbers. What results when dividing denominate numbers is either a constant value or a fractional rate. When the unit of measure is the same unit, you get a constant value. For example, 6 ft. divided by 3 ft. = 3ft. = $\frac{6\text{ft.}}{3\text{ft.}} = 2$. The unit of measure cancels out which leaves $\frac{6}{3} = 2$.

Fractional rates, or division by quantities, occur in real life all the time, especially in the industrial profession. For example, 10 mi. divided by 1 gal. results in the fractional rate 10 mi./1 gal. Other examples are 2,000 people/1 yr, slices/1 pizza, and 20 in./5 min.

As for other operations . . .

You can only add (or subtract) ***LIKE*** quantities, that is, only denominate numbers in the same unit of measure. But any quantities, like or not, can be multiplied together or divided. Remember, when you multiply quantities, you must multiply BOTH the numerical values and the unit of measures. However, when multiplying unlike quantities, it is often more practical, when possible, to convert to like quantities, and create a new quantity raised to power(s).

Addition/Subtraction: For example, to add 12 ft. and 6 in., you must convert one of the quantities to either all feet or all inches, so 6 in. is 0.5 ft., so 12 ft. + 6 in. = 12 ft. + 0.5 in. = 12.5 ft.

Or 12 ft. is 144 in., so 12 ft. + 6 in. = 144 in. + 6 in. = 150 in.

Multiplication: To multiply 12 ft. by 6 in., you can get (12 ft.) (6 in.) = 72 "ft. in." This may fit a purpose in answering some question that uses "feet inches," but a more practical measure would be converting inches to feet then multiplying.

As 6 in. = 0.5 ft., (12 ft.) (6in.) = (12 ft.) (0.5 ft.) = 6 ft · ft. = $6\,\text{ft.}^2$ (or 6 sq. ft.), a more common measure for area.

And we could also figure as 12 ft. = 144 in., then (12 ft.) (6 in.) = (144 in.) (6 in.) = 864 in. · 864 in. = 864 in^2 (or 864 sq. in.).

FRACTION OF A QUANTITY

Taking a fraction "of" a quantity means we multiply the fraction times the quantity.

Example: 1/3 of $27 = \frac{1}{3} \cdot 27 = \frac{1}{3} \cdot \frac{27}{1} = \frac{1 \cdot 27}{3 \cdot 1} = \frac{27}{3} = 9$

Example: 3/4 of $10\frac{6}{7} = \frac{3}{4} \cdot \frac{10 \cdot 7 + 6}{7} = \frac{3}{4} \cdot \frac{76}{7} = \frac{3 \cdot 76}{4 \cdot 7} = \frac{228}{28}$

$$\begin{array}{r} 8 \\ 28\overline{)\,228} \\ \underline{-224} \\ 4 \end{array}$$

$\frac{228}{28} = 8\frac{4}{28} = 8\frac{1}{7}$ so 3/4 of $10\frac{6}{7} = 8\frac{1}{7}$

Practice

$\frac{4}{5}$ of 70 ______, $\frac{4}{9}$ of 22 ______, $\frac{2}{11}$ of $1\frac{6}{7}$ ______.

Answers: 56; $9\frac{7}{9}$; $\frac{26}{77}$

ESTIMATION FRACTION OF QUANTITY

If we are taking a fraction of a mixed number, multiply the fraction time the closest whole number part to get the estimate of what the result should be.

Example: $\frac{2}{5}$ of $20\frac{1}{3}$ estimates 2/5 of 20 to get 8. So the result should be between 8 and 9.

When using two mixed number, round each to the closest whole number part, then multiply.

Example: $4\frac{1}{3}$ of $9\frac{6}{7} \cdot 4\frac{1}{3}$ is closest to 4, $9\frac{6}{7}$ is closest to 10, so four times $10 = 40$. So $4\frac{1}{3}$ of $9\frac{6}{7}$ is estimated to be about 40 (it is actually $42\frac{5}{7}$). Estimating is a good way to check to see if your exact answer makes sense.

1-3 Assignments

Multiply these fractions, simplify final fractions

1. $\frac{1}{9} \cdot \frac{5}{7} =$
2. $\frac{4}{9} \cdot \frac{7}{11} =$
3. $\frac{3}{7} \cdot \frac{1}{15} =$
4. $\frac{1}{12} \cdot \frac{1}{9} =$
5. $\frac{7}{8} \cdot \frac{6}{7} =$
6. $\frac{11}{20} \cdot \frac{11}{40} =$
7. $\frac{2}{5} \cdot \frac{1}{5} =$
8. $\frac{2}{5} \cdot \frac{5}{16} =$
9. $\frac{3}{4} \cdot \frac{14}{15} =$
10. $\frac{7}{8} \cdot \frac{7}{16} =$
11. $\frac{1}{2} \cdot \frac{3}{16} =$
12. $\frac{9}{20} \cdot \frac{1}{18} =$
13. $\frac{1}{b} \cdot \frac{5}{a} =$
14. $\frac{1}{xy} \cdot \frac{7}{2w} =$
15. $\frac{3t}{7} \cdot \frac{5t}{14} =$
16. $\frac{t}{p} \cdot \frac{y}{x} =$
17. $\frac{ab}{cd} \cdot \frac{ef}{gh} =$
18. $\frac{1}{f} \cdot \frac{h}{4f} =$
19. $2\frac{1}{7} \cdot 9\frac{5}{6} =$
20. $4 \cdot 1\frac{7}{12} =$
21. $7\frac{3}{10} \cdot 7\frac{1}{16} =$
22. $7\frac{1}{7} \cdot 1\frac{2}{9} =$
23. $5\frac{4}{9} \cdot 8\frac{3}{7} =$
24. $9\frac{1}{9} \cdot 6\frac{4}{41} =$
25. $8\frac{1}{2} \cdot 2\frac{2}{15} =$
26. $9\frac{8}{9} \cdot 6$
27. $10\frac{1}{12} \cdot 7\frac{3}{4} =$
28. $\frac{2}{3}$ *of* $500 =$
29. $\frac{7}{8}$ of $2121 =$
30. $\frac{1}{22}$ of $44000 =$
31. $\frac{4}{7}$ *of* $5\frac{1}{3} =$
32. $2\frac{5}{6}$ *of* $1100 =$
33. $1\frac{7}{8}$ *of* $4\frac{9}{13} =$

1-4 RECIPROCALS/DIVIDING FRACTIONS

RECIPROCAL

A reciprocal of a number is a value when multiplied the number to equal 1. We can easily turn any fraction into a reciprocal by "flipping" the numerator and denominator.

Example: The reciprocal of $\frac{3}{4}$ is $\frac{4}{3}$ and we know $\frac{3}{4}\cdot\frac{4}{3} = 1$

The reciprocal of $\frac{1}{9}$ is $\frac{9}{1}$ or 9 and we know that $\frac{1}{9}\cdot 9 = 1$.

The reciprocal of 11 is $\frac{1}{11}$ (note 11 can be written in fraction form as $\frac{11}{1}$).
Given a mixed number, simply convert it to improper form and then do your "flipping."

Example to find the reciprocal of $9\frac{4}{7}$ make $9\frac{4}{7} = \frac{9\cdot 7+4}{7} = \frac{67}{7}$ so the reciprocal of $9\frac{4}{7}$ is $\frac{7}{67}$.

DIVISION BY FRACTION

When one fraction (or number) is divided by a second fraction, find the second fraction's reciprocal, then multiply it by the first fraction (or number).

Example: $\frac{2}{7} \div \frac{5}{6}$ $\frac{6}{5}$ is the reciprocal of $\frac{5}{6}$ so we multiply $\frac{2}{7}\cdot\frac{6}{5} = \frac{12}{35}$

Example: $5 \div \frac{3}{4} = 5\cdot\frac{4}{3} = \frac{20}{3} = 6\frac{2}{3}$

Example: $\dfrac{\frac{2}{3}}{\frac{9}{11}} = \frac{2}{3} \div \frac{9}{11} = \frac{2}{3}\cdot\frac{11}{9} = \frac{22}{27}$

For division by a mixed number, convert to improper fraction and apply division rule.

Example: $5\frac{1}{2} \div 7\frac{1}{3}$ becomes $\frac{11}{2} \div \frac{22}{3} = \frac{11}{2}\cdot\frac{3}{22} = \frac{33}{44} = \frac{3}{4}$

Practice

$\frac{3}{7} \div \frac{5}{16}$ ________, $4 \div 2\frac{5}{6}$ ________, $4\frac{2}{3} \div 1\frac{9}{11}$ ________.

Answers: $1\frac{13}{35}$; $1\frac{7}{17}$; $2\frac{17}{30}$

ESTIMATION OF DIVISION OF FRACTIONS

Estimating division of fractions can be tricky. One technique is to round fractions to the nearest whole value, then divide the whole numbers to find a good estimate.

Example: Estimate $5\frac{3}{4} \div 2\frac{1}{3} \cdot 5\frac{3}{4}$ rounds to 6 and $2\frac{1}{3}$ rounds to 2, so $6 \div 2 = 3$. So $5\frac{3}{4} \div 2\frac{1}{3}$ is about 3 (it is actually $2\frac{13}{28}$)

Note: The problem is when a small value fraction, like ¼ or 1/11, which rounds to 0. So would require division.

1-4 Assignments

Divide these fractions, simplify final fractions

1. $\frac{1}{9} \div \frac{2}{3} =$
2. $\frac{4}{21} \div \frac{2}{7} =$
3. $\frac{3}{7} \div \frac{1}{15} =$
4. $\frac{1}{12} \div \frac{1}{8} =$
5. $\frac{7}{8} \div \frac{6}{7} =$
6. $\frac{8}{20} \div \frac{11}{30} =$
7. $\frac{2}{5} \div \frac{1}{5} =$
8. $\frac{2}{5} \div \frac{15}{16} =$
9. $\frac{3}{13} \div \frac{14}{15} =$
10. $\frac{7}{8} \div \frac{7}{16} =$
11. $\frac{1}{2} \div \frac{7}{16} =$
12. $\frac{9}{20} \div \frac{1}{12} =$
13. $2\frac{1}{7} \div 4\frac{1}{6} =$
14. $8 \div 1\frac{7}{12} =$
15. $14\frac{3}{10} \div 7\frac{1}{16} =$
16. $7\frac{4}{7} \div 5\frac{8}{9} =$
17. $8\frac{4}{9} \div 8\frac{3}{7} =$
18. $19\frac{7}{10} \div 6\frac{3}{5} =$
19. $8\frac{1}{2} \div 20\frac{5}{14} =$
20. $9\frac{2}{9} \div 6 =$
21. $2\frac{1}{16} \div 7\frac{3}{4} =$

1-5 POWERS AND ROOTS

POWERS

Exponents: Exponents are used to more easily write repeated multiplication of a number.

When we have $\boldsymbol{b^x}$, we have b, the base, and x, the exponent. This means we multiply b times itself x times. And we say "b raise to the power of x."

Example: 3^4; 3 is the base, 4 is the exponent, and this is 3 raised to the power of 4.

So, we multiply 3 by itself four times. Or $3^4 = 3 \cdot 3 \cdot 3 \cdot 3 = 81$

$$x^7 = x \cdot x \cdot x \cdot x \cdot x \cdot x \cdot x$$

When the exponent is 2, we have a special name for this operation. We say we "SQUARE" the base number, so, for example, 10^2 is 10 squared $= 10 \cdot 10 = 100$

When the exponent is 3, we say we "CUBE" the base number, so, for example, 4^3 is 4 cubed $= 4 \cdot 4 \cdot 4 = 64$.

USING THE CALCULATOR

On the scientific calculator, we use the ^ button (called the POWER key) when we want to apply an exponent. We enter the base number, followed by ^, then the exponent value, then ENTER.

So, to enter 5^3, we hit 5 ^ 3 ENTER to calculate (answer is 125).

For powers of 2, or squares, there is a special x^2 button you can use. You enter the base number followed by x^2 then ENTER and it will calculate the square of the number.

So, to find the square of 6, we enter 6 x^2 ENTER to calculate (answer is 36).

Practice

Use your calculator to find 7^4 _________ 13^3 ___________ 2.11^5 __________

Answers: 2401; 2197; 41.8227...

ESTIMATING POWERS

You can estimate exponential values by rounding to an easier number, then apply the repeated multiplication.

For example, 31^2 can be estimated as 30^2 or $30 \cdot 30 = 900$. So, 31^2 is just more than 900, (it is actually 961.)

ROOTS

A root is the inverse operation of a power, so, it "undoes" a power. Where $5^2 = 25$, we say 5 is the second root of 25, since five times $5 = 25$. When we take a second root, we also call this a "square root." So, 5 is the "square root" of 25. We utilize a sign, $\sqrt{\ }$, called a radical sign, to represent a square root. So, $5 = \sqrt{25}$.

Some square roots are easy to calculate—if we are finding roots of "perfect Square." Perfect squares are whole numbers that are squared, and it is useful to memorize as many perfect squares as one can. So, taking the square root of perfect squares is the whole number itself. For example, as $1^2 = 1$, then $\sqrt{1} = 1$, and $2^2 = 4$, then $\sqrt{4} = 2$, and so on.

Here are perfect squares for the first 30 whole numbers.

Chart of Perfect Squares 1 to 30		
$1^2 = 1$	$11^2 = 121$	$21^2 = 441$
$2^2 = 4$	$12^2 = 144$	$22^2 = 484$
$3^2 = 9$	$13^2 = 169$	$23^2 = 529$
$4^2 = 16$	$14^2 = 196$	$24^2 = 576$
$5^2 = 25$	$15^2 = 225$	$25^2 = 625$
$6^2 = 36$	$16^2 = 256$	$26^2 = 676$
$7^2 = 49$	$17^2 = 289$	$27^2 = 729$
$8^2 = 64$	$18^2 = 324$	$28^2 = 784$
$9^2 = 81$	$19^2 = 361$	$29^2 = 841$
$10^2 = 100$	$20^2 = 400$	$30^2 = 900$

USING THE CALCULATOR

For any number, the scientific calculator can calculate a square root in two ways:

A) Hit 2nd then x^2 followed by the number followed by ENTER. If not a perfect square, you will get a decimal answer.

B) You can enter the number, followed by ^ key then 0.5 (or [1 ÷ 2]). THIS IS BECAUSE TAKING A SQUARE ROOT IS THE SAME AS RAISING A NUMBER TO THE ½ EXPONENT POWER. So, $\sqrt{22} = 22^{.5} = 22^{\frac{1}{2}}$ and $\sqrt{1254} = 1254^{.5} = 1254^{\frac{1}{2}}$, and so on.

So to find the square root of 121, we enter 2nd x^2 121 ENTER (answer is 11).

Or we enter 121 ^ 0.5 ENTER or 121 ^ (1 ÷ 2) ENTER.

Practice

Find square roots for 169 ___________, 5625 ___________, and 455 ___________.

Answers: 13; 75; 21.3307...

CUBE ROOT AND OTHER ROOTS

Taking the third root is the inverse operation of raising to the third power.

If $4^3 = 64$, then the third root of 64 is 4. We call a third root the “cube root” of the number. We use the radical $\sqrt[3]{\ }$, So the cube root of 64 is written as $\sqrt[3]{64}$ (Note: Only with square roots, we do not place a number at the top of the radical sign.)

Just like square roots knowing perfect cubes makes calculating cube roots easy. Here are a few perfect cubes

$1^3 = 1$	$5^3 = 125$	$9^3 = 729$
$2^3 = 8$	$6^3 = 216$	$10^3 = 1000$
$3^3 = 27$	$7^3 = 343$	$11^3 = 1331$
$4^3 = 64$	$8^3 = 512$	$12^3 = 1728$

USING THE CALCUATOR

For any number, then the scientific calculator can calculate a cube root in two ways:

A) Enter 3 then 2nd, then ^ key, then the number, followed by ENTER. If not a perfect cube, you will get a decimal answer.

B) You can enter the number, followed by ^ key then (1 ÷ 3). THIS IS BECAUSE TAKING A CUBE ROOT IS THE SAME AS RAISING A NUMBER TO THE 1/3 EXPONENT POWER.

So $\sqrt[3]{234} = 234^{\frac{1}{3}}$ and $\sqrt[3]{76} = 76^{\frac{1}{3}}$, and so on.

Practice

Find the cube roots for 345 ______________, 216 __________, and 78.9 __________.

Answers: 7.0136..; 6; 4.2890..

OTHER ROOTS: Just as we can raise numbers to powers of 4, 5, 6, and so on, we can undo these by taking fourth, fifth, sixth, and on roots. There is no special names for these roots. We use the radical sign with the root number at the top, such as $\sqrt[4]{\ }$, $\sqrt[5]{\ }$, $\sqrt[6]{\ }$, and so on.

Use the calculator to find these roots, two ways

A) Type in the root number, then 2nd, then ^ key, then the number, followed by ENTER

B) You can enter the number, followed by ^ key then (1 ÷ root number) make sure the fraction is in parentheses when entering in calculator!

So $\sqrt[4]{23} = 23^{\frac{1}{4}}$ and $\sqrt[5]{12311} = 12311^{\frac{1}{5}}$, and so on.

ESTIMATING SQUARE ROOTS

Knowing the perfect squares can help you estimate a square root.

For example, take the square root of 10, or $\sqrt{10}$. Think of perfect squares that "sandwich" 10. That would be 9 and 16. We know the square root of 9 is 3, and the square root of 16 is 4. So the square root of 10 is between 3 and 4, and as 10 is just a little more than 9, we can estimate that $\sqrt{10}$ is a little more than 3 ($\sqrt{10}$ actually = 3.16 . . .).

Estimate $\sqrt{130}$. 130 is between perfect squares 121 and 144. It is also about halfway between 121 and 144. And since $\sqrt{121} = 11$ and $\sqrt{144} = 12$, then $\sqrt{130}$ should be about halfway between 11 and 12, say 11.5 ($\sqrt{130}$ is actually 11.402...).

Practice

Use estimation to find the square root $\sqrt{67}$ ______________.

1-5 Assignments

Find the following to the nearest hundredth.

1. 19^2 ________________
2. 3^7 ____________
3. 12^3 ________________
4. 4^4 ____________________
5. 2.5^6 __________________
6. 405^2 ___________________
7. 2.344^4 _____________
8. $(-5)^2$ __________________
9. $(-7)^3$ _______________
10. $36^{\frac{1}{2}}$ _______________
11. $8^{\frac{1}{3}}$ ___________________
12. $81^{\frac{1}{4}}$ _______________
13. $\sqrt{64}$ _________________
14. $\sqrt{225}$ ________________
15. $\sqrt{210}$ ______________
16. $\sqrt{10045}$ ______________
17. $\sqrt{.0041}$ ________________

18. $\sqrt[3]{216}$ ____________________

19. $\sqrt[3]{2344}$ ________________

20. $\sqrt[4]{345}$ ___________________

21. $\sqrt[5]{3000}$ _____________________

22. $\sqrt[6]{3412}$ ________________

23. $\sqrt[11]{2189}$ ____________________

24. $\sqrt[18]{65999}$ _________________

1-6 ORDER OF OPERATIONS

Order of Operations is a set of rules that dictate what operations to do first, then seconds, and so on, until we arrive at a single number, or simplified algebraic expression.

When expressions have more than one operation, we follow

1. First do all operations in proper order (items 2–4) to expressions that lie inside
   - Parentheses or brackets ( ), [ ], { }. Note: work innermost out.
   - Under radical signs $\sqrt{\ }$, $\sqrt[3]{\ }$, $\sqrt[4]{\ }$, and so on.
   - In numerators/denominator.
   - Inside absolute value.
2. Next, do any work, left to right, with exponents or radicals.
3. Again, working from left to right, do all multiplication or division, whatever comes first.
4. Finally, working from left to right, do all addition or subtraction, whatever comes first.

PEM/DA/S is an acronym used to remember this order of operation.

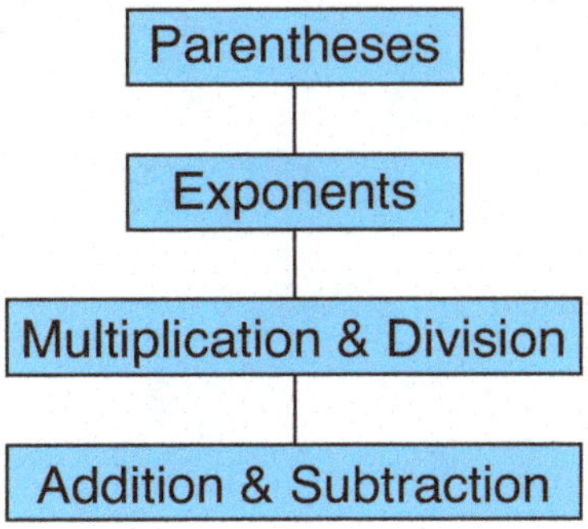

Practice

$$100 - 12 \div 3 + 32 \cdot 4 + 2$$

No parenthesis, no exponents (radicals, absolute value . . .) so look for multiplication/division, left to right.

$$100 - (12 \div 3) + 32 \cdot 4 + 2$$

Do division first, $12 \div 3 = 4$

$$100 - 4 + (32 \cdot 4) + 2$$

Do multiplication $32 \cdot 4 = 128$

$$(100 - 4) + 128 + 2$$

No more multiplication/division, look for addition or subtraction left to right, do $100 - 4 = 96$

$$(96 + 128) + 2$$

Do addition $96 + 128 = 224$

$$(224 + 2)$$

Do addition $224 + 2 = 226$

$$\text{so, } 100 - 12 \div 8 - 32 \cdot 4 + 2 = 226$$

Here's another one . . .

$$5^2 - 3[4 - 2\,(6 - 3)] \div 2$$

Do innermost parenthesis first

$$5^2 - 3[4 - 2\,((6 - 3))] \div 2$$

so do subtraction $6 - 3 = 3$

$$5^2 - 3[4 - 2\,(3)] \div 2 \text{ or } 5^2 - 3[4 - (2 \cdot 3)] \div 2$$

Now do [ ] brackets, multiply $2 \cdot 3 = 6$

$$5^2 - 3[(4 - 6)] \div 2$$

do subtraction $4 - 6 = -2$

$$5^2 - 3[-2] \div 2 \text{ or } (5^2) - 3 \cdot -2 \div 2$$

no more parentheses, so do exponent $5^2 = 25$

$$25\ (-3 \cdot -2) \div 2$$

No more exponents so do multiplication $3 \cdot -2 = -6$

$$25 - -(6 \div 2)$$

do division $-6 \div 2 = -3$

$$(25 - -3)$$

No more multiplication/division so do subtraction $25 - -3 = 25 + 3 = 28$

$$\text{so } 5^2 - 3[4 - 2(6 - 3)] \div 2 = 28$$

Example: Let's try this more complicated example!!!!! simplify

$$12 - [4 \cdot (3^4 \div 9 + 6 \cdot \frac{6+4}{18+2})] + \sqrt{8+92}$$

$$12 - [4 \cdot (3^4 \div 9 + 6 \cdot \frac{6+4}{18+2})] + \sqrt{8+92}$$

Start with innermost parentheses because numerators and denominators of fractions are groupings, do numerator/denominator first, $\frac{6+4}{18+2} = \frac{10}{20} = \frac{1}{2}$

$$12 - [4 \cdot (3^4 \div 9 + 6 \cdot \frac{1}{2})] + \sqrt{8+92} \text{ next do exponent } 3^4 = 81$$

$$12 - [4 \cdot (81 \div 9 + 6 \cdot \frac{1}{2})] + \sqrt{8+92} \text{ do division } 81 \div 9 = 9$$

$$12 - [4 \cdot (9 + 6 \cdot \frac{1}{2})] + \sqrt{8+92} \text{ do multiplication } 6 \cdot \frac{1}{2} = \frac{6}{2} = 3$$

$$12 - [4 \cdot (9 + 3)] + \sqrt{8+92} \text{ do addition } 9 + 3 = 12, \text{ drop inner parentheses}$$

$$12 - [4 \cdot 12] + \sqrt{8+92}$$

Do [ ] bracket, do multiplication $4 \cdot 12 = 48$ and remove brackets

$$12 - 48 + \sqrt{8+92} \text{ as under a radical is a grouping, do addition } 8 + 92 = 100$$

$$12 - 48 + \sqrt{100} \text{ do square root } \sqrt{100} = 10$$

$$12 - 48 + 10 \text{ do subtraction } 12 - 48 = -36$$

$$-36 + 10 \text{ do addition } -36 + 10 = -26$$

$$\text{so } 12 - [4 \cdot (3^4 \div 9 + 6 \cdot \frac{6+4}{18+2})] + \sqrt{8+92} = -26$$

Practice

1. $8 - 2 \cdot 4 - 12 \div 2 + 4 =$
2. $19 - [6 + 2(16 \div 8 \cdot 5 + 4)] \div 2 =$
3. $82 - 10\left(8 - \frac{6 + 2 \cdot 4}{18 - 11}\right)^2 \div \sqrt{25} =$
4. $12 \div 2 + 2[8 - (2 + 7) - \sqrt{4+12}]^2 =$

Answers: 1) −2, 2) 2, 3) 10, 4) 56.

1-6 Assignments

Evaluate these expressions using proper order of operation

1. $4(6) + 3 - 7(4) \div 2 =$
2. $5 \div 5 + 3 - 6 \cdot 2 =$
3. $3(7) - 11 + 15 \div 3 =$
4. $3 + 34 \div 17 + 4 \cdot 12 - 8 =$
5. $25 - (7 - 12 \div 6)(4) =$
6. $5[4 + (3)(2)] - (9 - 28 \div 7) \div 5 =$
7. $2^4 - 5(10 - 4^2 \div 2) + (30 - 3^3) =$
8. $(32 - 3^3 \div 9 \cdot 10)^5 - 4^2 \div 8 + 3^2$
9. $31 - \sqrt{36 \div 6 + 19} \llcorner 2^2 - 11 =$
10. $3^3 + \dfrac{\sqrt{3+2 \cdot 3+27}}{4-1} - 4(5) =$
11. $4^2 + (14 - 48 \div \dfrac{8+6 \cdot 4}{1+7}) - 7 \times 2 =$
12. $\dfrac{6 + 3 \cdot 5 - 6^2 \div \left[2 + \left(21 \div 3\right)\right]}{2^3 - 3 \cdot 2} =$

1-7 SCIENTIFIC CALCULATOR AND ORDER OF OPERATIONS.

NOTE: This course recommends the use of the scientific calculator, TI-30 xIIS. Demonstrations/ Instructions will be done using this calculator.

In the Industrial profession, it is critical that we have an understanding of how to enter calculations into our calculator. And it is important when we do any complicated calculation in the calculator we use proper order of operations, which means keeping the proper grouping in mind by utilizing the parentheses keys. Here are some helpful hints when using the calculator.

- If we want to square a negative number, say –4 to the 2nd power, we must remember to enter it as $(-4)^2$ and NOT as -4^2. Remember, when we multiply –4 by –4 we end up with a positive, as negative times negative is positive. So $(-4)^2$ will calculate the square of the negative and get the correct value, +16, whereas -4^2 will only square the 4, and will give an answer of –16.
- A mixed number needs to be treated like a grouped sum in the calculator. $3\frac{1}{2}$ would be written as $(3 + 1 \div 2)$

- Sums or differences in numerator or denominator. Let's say we have a fraction $\frac{2}{21+345}$ and we want to calculate it to a value on the calculator. Remember that the 21 + 345 in the denominator is a grouping and needs to be put in parenthesis. So it must be entered as 2 ÷ (21 + 345) which equals 2/366 ≈ .00546.

  NOTE: If it is entered in the calculator as 2 ÷ 21 + 345, without the parentheses, the calculator will follow order of operation, and the resulting calculation is to divide 2 by 21 then add that to 345, and get ≈ 345.0952

  The same thing goes for a sums or difference s in numerators For $\frac{23+567}{7}$ it must be entered as (23 + 567) ÷ 7 ≈ 84.29

- You need to practice with the calculator to become competent when to enter operations exact or when things need to be grouped. Examples:

  What about something like $3000\,(1 + .004 \cdot \frac{1}{4})^{20}$

  You could calculate the .004 times ¼ first to get .001, then add it to 1,

- Entering $3000\,(1 + .001)^{20}$ ≈ 3060. 5734 . . . to get the answer. However, you can enter your calculation steps as they come in one continuous flow, as long as you recognize proper grouping

  You can do so By entering 3000 (1 + .004 × 1 ÷ 4) ^ 20 which also yields about ≈ 3060. 5734 . . .

  In addition, here is a tip to make thing easier. As multiplying by 1 / 4 is really dividing by 4, this calculation can be done by dropping the X 1 and calculated as

  3000 (1 + .004 ÷ 4) ^ 20

  When we come to multiplying factors which are all fractions, such as $\frac{3}{4} \cdot \frac{5}{7} \cdot \frac{1}{8}$ you can do so in two ways

  1) we can put each fraction in parenthesis, then multiply, entering

     (3 ÷ 4) (5 ÷ 7) (1 ÷ 8) ≈ .0670

     Or

  2) Remember when we multiply fractions, we multiply the numerators over multiplying the denominators, so we can use the numerator product over the denominator product, making sure we group them in parentheses, like,

     (3 x 5 x 1) ÷ (4 x 7 x 8) ≈ .0670

- What about roots?

  Lets say we want to calculate $\sqrt{8 + 22^4}$.

  You could calculate $8 + 22^4$ first. As 22 ^ 4 = 234256, so $8 + 22^4$ = 8 + 234256 = 234264

  Then for $\sqrt{234256}$, you would do the square root either as 2nd x^2 234264 or as 234264 ^ .5 or as 234264 ^ (1 ÷ 2) each which would give you ≈ 484.01

  Or you can do it in one continuous flow using the operations as they come using parentheses.

This becomes Either as 2nd x^2 (8 + 22 ^ 4) ≈ 484.01

or as (8 + 22 ^ 4) ^ .5 ≈ 484.01

or as (8 + 22 ^ 4) ^ (1 / 2) ≈ 484.01

It is important what and when to put values in parentheses. And it comes from understanding order of operation and practice, practice, practice

PRACTICE, use scientific calculator and order of operation to calculate (so give answer in decimal to nearest hundredth):

a) $10\frac{3}{7} + 4\frac{2}{3}\left(6\frac{1}{6} - 5\frac{2}{3}\right)$

b) $\frac{4^2 + 5 \cdot (11 - 2^3)}{7 \cdot (-3)^2} + 6\frac{3}{11}$

c) $\sqrt{20^2 + 6\left(3\frac{1}{5}\right)}$

d) $\frac{3}{7} \cdot \sqrt{5 + 11^3} - \frac{(-26)^3 - 5^2}{27^3 + 8}$

Answers:

a) 12.76

b) 6.76

c) 20.47

d) 16.56

CHAPTER 2

Algebra: Expressions, Equations, and Formulas

2-1 VARIABLE EXPRESSIONS

In Mathematics, an **EXPRESSION** is a set of numbers and/or variables and operation signs that can set on one side of an equal sign.

Here are examples of expression that contain variable(s).

$$3x+2 \qquad -12x^2 5x-2 \qquad \sqrt{5y} \qquad \frac{2y+8}{y} \qquad \frac{23yxp+3x}{43} \qquad \log 23x+4z$$

An expression on its own with a **variable (or variables)** means, in most cases, that the unknown variable **COULD BE ANY REAL NUMBER VALUE**. For example, for $3x + 2$, x can be substituted with any real number value.

There are Exceptions. The value for a variable in some expressions can be limited when we are not able to do operations with certain numbers. For example, a single variable in a denominator can never be 0 since we cannot divide by 0. Another case is a variable under a radical sign for an even root can never be a negative.

EXAMPLE: given the expression $\frac{2y+8}{y}$, y can be any real number value except 0. We can see that $\frac{2y+8}{y} = \frac{2(0)+8}{0} = \frac{8}{0}$ which is undefined.

EXAMPLE: $\sqrt{5y}$ y can be any real number greater than or equal to 0. You can see if $y < 0$, then $5y$ would be negative, and you cannot have a negative under a square root radical.

Practice

Identify what values for the variable can be used in these expressions

a) $15x+29$ __________

b) $2y^2+35y-2$ __________

c) $\sqrt{w-5}$ __________

d) $\frac{2t+18}{t}$ ____________

e) $16h^{.5}$ ____________

f) $5(3x+2)-7(x+7)$ ____________

g) $\frac{5}{3-y}$ ____________

Answers: a) all real numbers, b) all real numbers c) w > or equal to 5 d) t not equal to 0; e) h > or equal to 0, f) all real numbers g) y not equal to 3

2-1 Assignments

Identify what values for the variable can be used in these expressions. If any real number can satisfy the expression, write ALL REAL NUMBERS. If the set of values is limited, describe what that set is.

1. $32x$ ____________
2. $y^2 + y - 12$ ____________
3. $5x + 45$ ____________
4. $4t^3 - 6t^2 - 3t + 4$ ____________
5. $\sqrt{w}$ ____________
6. $\sqrt{p+8}$ ____________
7. $\frac{22}{x-7}$ ____________
8. $\frac{5x+7}{3x+8}$ ____________
9. $(x+8)-6(x+66)$ ____________
10. $\frac{x^2+3x-4}{x-11}$ ____________

2-2 SIMPLIFYING VARIABLE EXPRESSION

An EQUATION is a statement consisting of making two expressions equal, or placing an = sign between the expressions. When we take two expressions with at least one of them with a variable (or variables) and make them equal, we can then try to solve for the variable values using properties of mathematic equality.

Before we can solve any variable equations, it is essential that both expression sides of the equation are SIMPLIFIED.

When we simplify expressions, we need to follow these rules through completion

1. remove parentheses by multiplying factors.
2. Combine like terms so that each variable to a specific power appear just once in the expression
3. combine the constants.
4. All constants and/or coefficients are simplified, including fractions.

SIMPLIFYING EXPRESSIONS TOOLS

Combining LIKE TERMS

LIKE TERMS are terms that have the same variable(s) to the same exponent power.

Examples of like terms are x, $4x$, $11x$, $.5x$, $-34x$. Note, each share a variable factor of x

Another set of like terms t^2, $4t^2$,$-t^2$, $10000t^2$. Note each share a variable factor of t^2

Another set of like terms are x^2y, $4x^2y$, $-.01x^2y$. Note each share a variable factor of x^2y

Also, numbers by themselves, called CONSTANTS, are consider like terms as well.

However, $4w$ and $10w^2$ are **not like terms**. They share a variable w, but each term has a different exponent value.

Other examples of terms that are not like are $2y^3$ *and* $9y^9$; 5 and x ; tr^3 *and* t^3r

ADDITION/SUBTRACTION RULE SAYS:

- We can add (or subtract) like terms and simplify.
- We cannot add (or subtract) terms that are not like to simplify. We need to leave the sum as it is.

HOW TO ADD (OR SUBTRACT) LIKE TERMS.

When the variable factor has a number in front of it, such as $3x$ or $7y^2$, we are multiplying the variable by that number. So $3x$ is 3 times x and $7y^2$ is 7 times y^2. The number in front is called the variable's COEFFICIENT. So for $3x$, 3 is the coefficient of x, For $7y^2$, 7 is the coefficient of y^2. When our term is just a variable, we are assuming multiplication by 1, so the coefficient is 1. So for the variable expression x, 1 is the coefficient.

Take the expression

$$4x^2 + 5x^2 + x^2$$

This expression is <u>not</u> simplified, as x^2 shows up 3 times in the expression.

We need to add, or combine, the like terms by adding the coefficients of each term and make that sum a single coefficient for our variable.

As 4 is the coefficient of $4x^2$, 5 is the coefficient of $5x^2$, and 1 is the coefficient of x^2, we can write this expression as

$$4x^2 + 5x^2 + x^2 = (4+5+1)x^2 = 10x^2$$

Here is an example of the difference of like terms:

$$10y^3 - 4y^3 = (10-4)\,y^3 = 6y^3$$

So the rule is

ADD/SUBTRACT LIKE TERMS BY ADDING/SUBTRACTING THE COEFFICIENTS AND KEEPING THE VARIABLE BASE AND ITS EXPONENT THE SAME.

Practice Simplifying these Expressions

a) $9a + 13a - 7a$ ______________

b) $13a + 9 + 9b + 12a - 7b$ ______________

c) $12z + 5 + 3z$ ______________

d) $10m + 12n - 5m + 3n + 5$ ______________

e) $4c^2 + 3c + 11 + 8c^2 - 9$ ______________

f) $13ab + 4b + 3ab - 2b^3 + 9a - 7b$ ______________

Answers: a) 15a; b) 25a + 2b + 9; c) 15z + 5; d) 5m+ 15n + 5; e) 12 c^2 + 3c +2; f) -2b^3 +16ab - 3b + 9a

DISTRIBUTION PROPERTY

We need to use the DISTRIBUTIVE PROPERTY sometimes to simplify an algebraic expression. This gets rid of parentheses in the expression.

The **DISTRIBUTIVE PROPERTY** says

$$x\,(y + z) = xy + xz \text{ and } x(y - z) = xy - xz$$

Here are Examples of the distributive property

$$3(x + 2) = 3(x) + 3(2) = 3x + 6$$

$$4(y - 2) = 4(y) + 4(-2) = 4y - 8$$

$$x(x+7) = x(x) + x(7) = x^2 + 7x$$

$$x(w + x - 10) = x(w) + x(x) + x(-10) = xw + x^2 - 10x$$

$$\frac{2}{3}\left(\frac{4}{7}x + \frac{1}{2}\right) = \frac{2}{3}\left(\frac{4}{7}x\right) + \frac{2}{3}\left(\frac{1}{2}\right) = \frac{2\cdot 4}{3\cdot 7}x + \frac{2\cdot 1}{3\cdot 2} = \frac{8}{21}x + \frac{2}{6} = \frac{8}{21}x + \frac{1}{3}$$

What about $-(x - 8)$?

This also employs the distributive property, as we are multiplying -1 times $(x - 8)$

$$-(x-8) \text{ is equal to } -1(x - 8) = -1(x) + -1(-8) = -x + 8.$$

Note: A quick way to simplify with multiplying by -1 in the distributive property is change the signs of the terms in the sum.

Example: $-1(d + 5) = -d - 5$ $-1(-w - 9) = w + 9$

Practice

a) $3(w + 11) =$ ____________

b) $5(k - 4) =$ ______________

c) $-12q(q - 9) =$ ________

d) $14(x + w + 3r) =$ _____________

e) $-123(-3x + 2.3y) =$ __________

f) $3t(t^2 - t + 2) =$ ____________

g) $\frac{1}{4}\left(\frac{5}{6}p + \frac{12}{5}\right) =$ ___________

h) $\frac{3}{4}x\left(b - \frac{4}{3}\right) =$ _________

Answers: a) 3w + 33; b) 5k – 20; c) $-12q^2 + 108q$; d) 14x + 14w + 42 r; e) 369x – 282.9 y; f) $3t^3 - 3t^2 + 6t$; g) 5p/24 + 3/5; h) 3xb/4 – 1

2-2 Assignments

Simplify these expressions

1. $19x + 12x - 7x$ ______________
2. $5y - 6y + 11y - 5y$ _______________
3. $34g + 12g - 5g + 200g$ ___________
4. $10f + 9 + 29h + 12f - 7 + 11h$ ____________
5. $12.5k + 5 + 3.23k - 5$ ________________
6. $10x^2 + 12x - 35x + 3x^2 + 5$ __________________
7. $24d^2 + 3c + 11d + 22c^2 - 9d + c^2$ _______________
8. $3ab + 6b + 13ab - 2a^3 + 19a - 17b$ __________
9. $100x^2 + 2x - 2 + 5x - 23x^2 + 5 - 77x^2 - 3$ __________
10. $7(k + 10) =$ ____________
11. $-(3k - 8)$ _____________
12. $25(3h - 4) =$ ______________
13. $-r(r + 22) =$ ________
14. $18(2x + 3w + 4y) =$ ______________

15. $2(-7x + 1.3y) =$ ___________

16. $6y(y^2 + 2y - 12) =$ ___________

17. $\frac{1}{3}\left(\frac{5}{7}h + \frac{2}{5}\right) =$ ___________

18. $\frac{3}{8}x\left(16x - \frac{8}{3}\right) =$ ___________

19. $6(3x + 2) - 9(4x - 8) + 13x$ ________________

20. $4t(k + 9h + p) - 6tk + 4th + 9tp$ _____________

21. $8(3m + 2) - 9(4n + 8) + 7(m + n)$ _____________

22. $7(4k + 6y - 10x) - 2(14k + 21y - 35x)$ _________

2-3 SOLVING SIMPLE VARIABLE EQUATIONS

Remember, we have to SIMPLIFY OUR ALGEBRAIC EXPRESSIONS in order to SOLVE ALGEBRAIC EQUATIONS. Here are the rules for simplifying expressions again.

Remove parentheses by multiplying factors.

Combine like terms so that each variable to a specific power appear just once in the expression

Combine the constants.

All constants and/or coefficients are simplified.

Therefore: An expression is simplified if

- there are no parentheses,
- all variables raised to a particular power appears only once,
- all constants are reduced to one constant

Only when we have simplified expressions on both sides of the equation can we then solve our variable in an equation.

SOLVING FOR A VARIABLE IN AN EQUATION

We need some tools to arrive at a solution. We can use the principle OF EQUALITY:

> WHATEVER YOU DO TO ONE SIDE OF THE EQUATION, YOU DO TO THE OTHER SIDE TO RETAIN EQUALITY.

So You can do the following to both sides of an equation

- Add
- Subtract
- Multiply

- Divide
- Raise to the same exponent power
- Take the square root, cube root, 4th root, ...
- Take the logarithm

And your equation remains EQUAL.

The goal is to do these operations in order to **ISOLATE YOUR VARIABLE** so it equals eventually a number value. And the strategy is to use inverse operations on both sides of the equation to undo operations already there.

Solve $11x + 6x = 204$

The left side is not simplified as the x appears twice. We need to combine terms on the left sides.

$11x + 6x = (11 + 6)x = 17x$

So $11x + 6x = 204$ becomes

$17x = 204$. Now both sides of the equation are simplified.

Now to isolate the x, we need to "undo" the multiplication of $17x$ by dividing both sides of the equation by 17

$$\begin{array}{rl} 17x = 204 & \\ \div 17 \quad \div 17 & \text{we get} \\ \hline x = 12 & \text{which is our solution} \end{array}$$

Solve $4(x + 5) - 2x + 9 + 2(x + 4) = 233$

We need to use the distributive property <u>twice</u> on the left side

$$4(x + 5) - 2x + 9 + 2(x + 4) = 4(x) + 4(5) - 2x + 9 + 2(x) + 2(4) = 4x + 20 - 2x + 9 + 2x + 8$$

We need to combine like terms so there is only one x and one constant on the left side.

We can rearrange our terms so that like terms are together

$$4x + 20 - 2x + 9 + 2x + 8 = 4x - 2x + 2x + 20 + 9 + 8,$$

now we can combine like terms

$$4x - 2x + 2x + 20 + 9 + 8 = (4 - 2 + 2)\,x + (20 + 9 + 8) = 4x + 37$$

So $4(x + 5) - 2x + 9 + 2(x + 4) = 233$ becomes $4x + 37 = 233$

We need to isolate the x on the left side. We can use subtraction then division to undo the operations

$$\begin{array}{rl} 4x + 37 = 233 & \text{subtract 37 both sides} \\ -37 \quad -37 & \\ \hline 4x = \ 196 & \text{divide both sides by 4} \\ \div 4 \qquad \div 4 & \\ \hline x = \ 196 & \text{which is our solution} \end{array}$$

What if there are variables on both sides of an equation? We still need to simplify each side first, then we use properties of equality to get numbers one side and the variable on the other side of the equation.

Solve $3(x + 2) = 7(x - 6)$, We need to simplify both sides.

We use the distributive property both sides, $3(x) + 3(2) = 7(x) + 7(-6)$ which becomes $3x + 6 = 7x - 42$

The equation is now simplified on both sides

$$\begin{array}{rcl} 3x + 6 & = & 7x - 42 \\ -6 & & -6 \\ \hline 3x & = & 7x - 48 \\ -7x & & -7x \\ \hline -4x & = & -48 \\ \div -4 & & \div -4 \\ \hline x & = & 12 \end{array}$$

we can subtract 6 from both sides

we can subtract $7x$ from both sides

we divide both sides by -4

our solution

Solve this one: $.25C + 3 = C$

Remember, when there is no number in front of our variable C, we assume that the coefficient is 1 and we are multiplying by C by 1.

$$\begin{array}{lcl} .25C + 3 & = & 1.C \\ -\ .25C & & -.25C \\ \hline 3 & = & (1 - .25)C \\ 3 & = & .75C \\ \div .75 & & \div .75 \\ \hline 4 & = & C \end{array}$$

both sides are simplified. So we can subtract $.25C$ from both sides

As $1 - .25 = .75$

Divide both sides by .75

our solution

Practice

a) $3x + 12 = 33$

b) $\frac{1}{2}(m + 4) = 49$

c) $3y + 8(y - 4) = -43$

d) $4p + 9p + 4(p - 6) = 19(p + 2)$

e) $.5x + 2(x - 4) = 4(x + .75)$

f) $9(g + 2) - 4g + 12 = 2(2g + 15)$

Answers: a) 7; b) 94; c) −1; d) −31; e) − 7.333...; f) 0

2-3 Assignments

Solve for the variable.

1. $4x = 48$
2. $5y - 1 = 19$
3. $32t + 100 = -60$
4. $5t + 43 = -68$
5. $23x + 14 = 111$
6. $2.5t + 3.5 = 12$
7. $.5(m - 6) = 40$
8. $11y + 10(y - 8) = 100$
9. $1.5x + 2.5(3x - 8) = 24$
10. $w + 7w - 4(w - 22) = 11(w + 3)$
11. $.5r + 6(r - 70) = 2(r + 1.75)$
12. $8(d + 8) + 7(d - 2) + 6(4d - 9) = 7(d - 1)$
13. $7(2g + 4) - 8g + 6 = 2(9g + 19)$
14. $7(11r - 4r) + 4 = 34r$
15. $8(p + 2) - 6(p - 5) + 8(p - 5) = 6(p - 2) + 8(p - 3) + 2(p + 6)$

2-4 RATIOS AND PROPORTIONS

Another name for a fraction is a RATIO. When one fraction equaling another fraction it is called a PROPORTION. For example, $\frac{1}{2} = \frac{5}{10}, \frac{13}{39} = \frac{1}{3}$, or $\frac{x}{3} = \frac{7}{35}$ are all proportions. We say the numerator to denominator changes in the same way for both fraction/ratios in proportions.

When we have ratios (or fractions) with a variable in our equation, we call these RATIONAL EQUATIONS. When the rational equation involves a proportion, we can solve it by using what is called **CROSS MULTIPLICATION**. It is also called the **MEANS/EXTREMES** property.

In any proportion, the means/extremes property says for

$$\text{if } \frac{a}{b} = \frac{c}{d}, \text{ then } ad = bc$$

The numerator a and denominator d are considered the "extremes" of the proportion, and the denominator b and numerator c are considered the "means" of the proportion. So given the proportion, the product of the extremes equals the product of the means.

You should see the "cross multiplying" in this property

$$\frac{a}{b} = \frac{c}{d} \text{ becomes } ad = bc$$

Solve $\frac{3}{5} = \frac{x}{10}$ for x we cross multiply to get $(3)(10) = (5)(x)$ or

$30 = 5x$ divide both sides by 5

$\div 5 \quad \div 5$

$6 = x$ our solution

So $\frac{3}{5} = \frac{6}{10}$

Example: Solve for y $\frac{3}{y} = \frac{11}{430}$ cross multiply $430(3) = y(11)$

$1290 = 11y$ divide both sides by 11

$\div 11 \quad \div 11$

$\frac{1290}{11} = y$

Practice solve for variable **a)** $\frac{2}{3} = \frac{w}{50}$ **b)** $\frac{7}{11} = \frac{2}{h}$

c) $\frac{2q}{5} = \frac{4}{9}$ **d)** $\frac{5}{9.6} = \frac{2}{4.5x}$

Answers: a) 100/3; b) 22/7; c) 10/9; d) .8533...

WHAT IF YOU HAVE A SUM OR DIFFERENCE IN A NUMERATOR OR DENOMINATOR IN A TWO RATIONAL EXPRESSION EQUATION

Example: $\frac{x+3}{x} = \frac{3}{4}$ Because we have single term fractions (and the $x + 3$ forms a single value in the numerator), we, therefore, have a proportion, and we can use cross multiplication to solve for x.

$\frac{x+3}{x} = \frac{3}{4}$ becomes $4(x + 3) = 3(x)$ using distributive property on the left, we get

$4x + 12 = 3x$ subtract $4x$ from both sides

$-4x \quad -4x$

$12 = -x$ divide both sides by -1

$\div -1 \quad \div -1$

$-12 = x$

$\frac{4}{h+3} = \frac{11}{2h-3}$ This is a proportion, so cross multiply

$\frac{4}{h+3} = \frac{11}{2h-3}$ becomes $4(2h-3) = 11(h+3)$ distribute both sides

$8h - 12 = 11h + 33$ add 12 both sides

$\underline{+12 \quad +12}$

$8h = 11h + 45$ subtract $11h$ both sides

$\underline{-11h \, -11h}$

$-3h = 45$ divide both sides by -3

$\underline{\div -3 \quad \div -3}$

$h = -15$

Practice

a) $\frac{x}{x+1} = \frac{13}{15}$

b) $\frac{p-5}{2} = \frac{p}{200}$

c) $\frac{r+2}{5} = \frac{r-10}{4}$

d) $\frac{m-5}{7-m} = \frac{9}{13}$

Answers: a) 13/2 ; b) 500/99 ; c) 58; d) 64/11

REMEMBER YOU CANNOT CANCEL ACROSS A SUM OR DIFFERENCE IN A FRACTION

we know that $\frac{24+8}{8} = \frac{32}{8} = 4$

This is not the same as $\frac{24+\cancel{8}}{\cancel{8}} = \frac{24+1}{1} = 25.$

You cannot cancel out the 8s across the sum in the numerator.

For $\frac{20x+11}{5}$, you cannot cancel the 5 out of just the $20x$ to get $\frac{4x+11}{1}$. The only way you can cancel out the 5 is if 5 is a common factor of both terms in the numerator. Because 5 is not a factor of 11, the best we can do is to keep it as $\frac{20x+11}{5}$. It is simplified!

If you had $\frac{20x+15}{5}$, the 5 is a factor of both $20x$ and 15, so we can cancel out a 5 in each term to get $\frac{20x+15}{5} = 4x+3$. So it is now simplified.

YOU CAN ONLY CROSS MULTIPLY WHEN YOU HAVE TWO TERMS IN A RATIONAL EQUATION

So . . . You cannot simply cross multiply when you have more than two terms in a rational equation.

For example, $\frac{y}{2} + 8 = \frac{4}{11}$ has three terms, so you cannot cross multiply as is. However, you can change this into a two term proportion and then use cross multiplication. This will be discussed in Section 2-5.

We can use cross multiplication to solve proportion word problems involving conversions.

Example: Richie Rich is vacationing in Europe. He needs to convert \$450 to Euros. Richie checked the exchange rate and found it to be 1 Euro = \$1.23
We want to find how many Euros Richie will get. So first let x = number of Euros.

The relationship between Euros and Dollars is the ratio $\frac{1 \text{ Euro}}{\$1.23}$

We can now set up a proportion to find the equivalent Euro value for \$450.

We get $\frac{1 \text{ Euro}}{\$1.23} = \frac{x \text{ Euros}}{\$450.00}$ We can write this without units of measure as $\frac{1}{1.23} = \frac{x}{450}$ Now we cross multiply $(1)(450) = (1.23)(x)$ or $450 = 1.23x$

$$\frac{\div 1.23 \quad \div 1.23}{365.85 = x}$$

Rounding to the nearest hundredth of a Euro, \$450 = 365.85 Euros

Example: Mr. Magoo wants to go to Canada. He has \$2900.03 and needs to convert it to Canadian Dollars.

Currently the exchange ratio is 1.00 Canadian \$ = \$0.81 US

Let C = Canadian dollar amount

Set up exchange ratio $\frac{\$1.00 \text{ Canadian}}{\$0.81}$

Set it equal to the ratio $\frac{C \text{ \$ Canadian}}{\$2900.03}$ or $\frac{1.00}{0.81} = \frac{C}{2900.03}$

Solve $\frac{1.00}{0.81} = \frac{C}{2900.03}$ by cross multiplying

$$(1.00)(2900.03) = (0.81)(C)$$
$$2900.03 = 0.81C$$
$$\frac{\div 0.81 \quad \div 0.81}{3580.28 = \quad C}$$

So \$2900.03 is equal to \$3580.28 Canadian

Example: Wellington weights 30 kg. What is his weight in pounds?

1 kg is about 2.2 lb.

Let P = number of pounds.

Conversion ratio is $\frac{1 \text{ kg}}{2.2 \text{ lb.}}$, set it equal to $\frac{30 \text{ kg}}{P \text{ lb.}}$ or $\frac{1}{2.2} = \frac{30}{P}$

Solve $\frac{1}{2.2} = \frac{30}{P}$ by cross multiplying $(1)(P) = (2.2)(30)$ or $P = 66$

So 30 kg is equal to about 66 lb. Therefore, Wellington weight is about 66 lb.

Example: Fast Freddy drives his motor car at 88 mph. How many kilometers does he travel in 300 min?

We need to do two conversions, turning 88 mi. into km and turning 300 min into hr.

The conversion factor for miles to kilometers is 1 km = 0.62 mi.

Let k = number of kilometers

So $\frac{1\text{ km}}{0.62\text{ mi.}} = \frac{k\text{ km}}{88\text{ mi.}}$ or $\frac{1}{0.62} = \frac{k}{88}$

Solve $\frac{1}{0.62} = \frac{k}{88}$ by cross multiplying $(1)(88) = (0.62)(k)$

$$\begin{aligned} 88 &= 0.62k \\ \div 0.62 &\quad \div 0.62 \\ \hline 141.9354\ldots &= k \end{aligned}$$

Rounding to the nearest hundredth of a kilometer, 88 mi. = 141.94 km. So if Fast Freddy drives 88 mi. in 1 hr, then he drives 141.94 km in 1 hr.

Now we need to convert 300 min to hr

Let h = number of hours

Since there is 60 min in 1 hr, then $\frac{1\text{ hr}}{60\text{ min}} = \frac{h\text{ hr}}{300\text{ min}}$ or $\frac{1}{60} = \frac{h}{300}$

Solve $\frac{1}{60} = \frac{h}{300}$ by cross multiplying $(1)(300) = (60)(h)$

$$\begin{aligned} 300 &= 60\text{h} \\ \div 60 &\quad \div 60 \\ \hline .5 &= h \end{aligned}$$

So, 300 min = 5 hr.

We now need the number of kilometers driven in 5 hr, let x = kilometers driven in 5 hr, and we can say $\frac{141.94\text{ km}}{1\text{ hr}} = \frac{x\text{ km}}{5\text{ hr}}$ or $\frac{141.94}{1} = \frac{x}{5}$

Cross multiply $(141.94)(5) = 1(x)$

$$709.70 = x,$$

so Fast Freddy drive to get $709.70 = x$ km in 300 min

Practice

a) Quagmire has 420 g of mercury. He needs to convert this to ounces. 1 g = 0.04 oz.

b) Pep Cee has 123 fl. oz. of a tasty beverage. Pep needs to know how many liters of the beverage he has.
I L = 33.81 fl. oz.

c) Watson's Elementary Dairy sells milk for \$1.19 per 16 fl. oz.

What would 3.2 gal. of milk cost if this rate applies.

There are 128 fl. oz. in 1 gal.

Answers:

a) x ounces/420g =0.04oz/1g, x = 16.8 oz;

b) x liters/123 fl oz = 1 liter /33.81 fl oz, x = 3.64 liters;

c) 3.2 gal x 128 fl oz /1 gal = 409.60 fl oz , x / 409.60 fl oz = \$1.19/16 fl oz , x = \$30.46

2-4 Assignments

Solve for the variable

1. $\frac{5}{6} = \frac{w}{60}$ ________

2. $\frac{q}{12} = \frac{5}{9}$ ________

3. $\frac{y}{17} = \frac{23}{51}$ ________

4. $\frac{3}{7} = \frac{72}{f}$ ________

5. $\frac{7}{11.5} = \frac{6}{3.5x}$ ________

6. $\frac{9}{13} = \frac{3}{1.7h}$ ________

7. $\frac{3m}{7} = \frac{m-2}{31}$ ________

8. $\frac{x}{x+2} = \frac{19}{35}$ ________

9. $\frac{r-9}{3} = \frac{r}{100}$ ________

10. $\frac{5+2d}{9} = \frac{d-11}{4}$ ________

11. $\frac{k-5}{9-k} = \frac{8}{13}$ ________

12. $\frac{h+15}{h-20} = \frac{12}{13}$ ________

Solve the following using the conversion equation in proportions

13. Henry has 56 Liters of Lemonade. He needs to know how many quarts of Lemonade he has. 1.00 quart = .95 Liters.

14. Delores has 456 grams of conditioner. She needs to know how many ounces of conditioner she has. 1.00 ounce = 28.35 grams

15. Paula walked 4.5 miles. How many kilometers did Paula walk? 1 mile = 1.61 kilometers

16. The package has a volume of 460 cubic feet. How many cubic meters is the package? 1.0 cubic meter = 35.32 cubic feet

17. The new Howdy Doody postage stamp has an area of 2.08 in^2. What is the area of this stamp in cm^2? 1.0 $in^2 = 6.45\ cm^2$

18. Mom and Pop's Shopette ordered 910.5 pints of half and half cream. How many gallons of half and half did they order? 1.0 gallon = 8.0 pints

19. Speedie ran the 200 meter race in 25.5 seconds. How much of an hour did it take Speedie to run the race? 1.0 hour = 3600.0 seconds

20. A great whale weighed 9560.4 kilograms. What is the weight in tons? 1.0 ton = 907.03 kilograms

21. Wendell' Plumbing Supply sold metal pipping at $36.50 per 15 yard length. What would 75.4 meter of piping cost. 1.0 meter = 1.09 yards

22. It cost $2 per gallon of pool cleaner to keep the pool antiseptic. It takes 1 gallon of pool cleaner to properly clean 9 gallons of pool water. The swimming pool currently has 39600 cubic inches of water. a) How many gallons of water are in the pool? b) And how much does it cost for pool cleaner? 1 gallon = 231 cubic inches.

2-5 SOLVING RATIONAL EQUATION

In Section 2-4, we dealt with rational equations when you have a two term variable proportion, using means-extremes property, or cross multiplication to solve them. And it was noted that **YOU CAN ONLY CROSS MULTIPLY WHEN YOU HAVE TWO TERMS IN A RATIONAL EQUATION.**

What if you have more than two terms in a rational equation? We will show two methods to solve these: (A) CONVERTING INTO A PROPORTION and (B) MULTIPLYING BY A COMMON DENOMINATOR MULTIPLE.

A) CONVERTING INTO A PROPORTION

For Example: $\frac{y}{2} + 8 = \frac{4}{11}$

The goal is to get the $\frac{y}{2}$ on one side and a fraction on the other side. We can accomplish this by subtracting 8 both sides.

$$\frac{y}{2} + 8 = \frac{4}{11} \text{ subtracting 8 both sides}$$
$$\underline{-8 \quad -8}$$
$$\frac{y}{2} = \frac{4}{11} - 8 \text{ For the left side using the fraction subtraction rules,}$$

$$\frac{4}{11} - 8 = \frac{4}{11} - \frac{8}{1} = \frac{4}{11} - \frac{8 \cdot 11}{1 \cdot 11} = \frac{4}{11} - \frac{88}{11} = \frac{4-88}{11} = \frac{-84}{11}$$

So $\frac{y}{2} = \frac{-84}{11}$ Now we have a proportion, and we can cross multiply

$\frac{y}{2} = \frac{-84}{11}$ Becomes $11y = 2(-84)$ or $11y = -168$ divide both sides by 11

$$\underline{\div 11 \quad \div 11}$$
$$y = \frac{-168}{11}$$

Example: $\frac{5}{y} + \frac{1}{2} = \frac{13}{4}$ subtract $\frac{1}{2}$ both sides

$$\underline{-\frac{1}{2} \quad -\frac{1}{2}}$$
$$\frac{5}{y} = \frac{13}{4} - \frac{1}{2} = \frac{13}{4} - \frac{2}{4} = \frac{13-2}{4} = \frac{11}{4}$$

So $\frac{5}{y} = \frac{11}{4}$ Now we have a proportion, so we cross multiply

$5(4) = y(11)$

$20 = 11y$ divide both sides by 11

$\underline{\div 11 \div 11}$

$\frac{20}{11} = y$

Practice using conversion to proportion

a) $\frac{2x}{3} + \frac{5}{6} = \frac{10}{6}$

b) $\frac{5}{2y} + 2 = \frac{1}{2}$

c) $\frac{5x}{11} - \frac{7}{9} = \frac{1}{9}$

d) $1 + \frac{11}{3t} = 2\frac{12}{13}$

Answers: a) 2x/3 = 5/6, x = 15/12 or 5/4; b) 5/2y = -3/2, y = 10/-6 or y = -5/3; c) 5x/11 = 8/9, x = 88/45; d) 11/3t = 25/13, t = 143/75

B) MULTIPLYING BY A COMMON DENOMINATOR MULTIPLE

Another handy technique to solve a rational is one designed to eliminate all the denominators, so NO MORE FRACTION!!!! YEAH! This is accomplished by find a common multiple of each of the denominators and then multiplying both sides of your equation.

Here are the steps

- Simplify each fraction in our rational equations, if possible.
- Then multiply both sides of the equation by a COMMON MULTIPLE OF ALL THE DENOMINATORS. Ideally, the least common multiple (LCM) of the denominators is the best one to use.
- Then use algebra steps, factoring, and so on, to solve.

NOTE: If you don't know the LCM of the denominators, you can always find the product of all the denominators, and use that as your multiplier. It will cancel out the denominators, but may require simplifying your answer at the end.

Example:

Solve $\frac{x}{2}+\frac{x}{3}=10$

The denominators of our three terms are 2, 3, and 1.

The LCM of the denominators 2, 3, (and 1) is 6. So we want to multiply both sides of the equation by 6 to eliminate the denominators.

$6\left(\frac{x}{2}+\frac{x}{3}\right)=6(10)$ use distributive property right side

$6\left(\frac{x}{2}\right)+6\left(\frac{x}{3}\right)=6(10)$ multiply through both sides

$3x+2x=60$ Note: Fractions eliminated!! Now combine like terms right side

$5x=60$ divide both sides by 5

$x=12$

Another example: $\frac{x+2}{3}=\frac{x-4}{7}$

The LCM of 3 and 7 is 21, so multiply both sides by 21,

$$21\cdot\frac{x+2}{3}=21\cdot\frac{x-4}{7}$$

$7(x+2)=3(x-4)$ use distributive property both sides

$7x+14=3x-12$ subtract 14 both sides

$7x=3x-26$ subtract $3x$ both sides

$4x=-26$ divide 4 both sides

$x=-\frac{26}{4}=-\frac{13}{2}$

Example: solve for x $\frac{2}{3} + \frac{3}{4x} = \frac{7}{12}$ the LCM of 3, $4x$, and 12 is $12x$

So multiply both sides by $12x$

$$12x\left(\frac{2}{3} + \frac{3}{4x}\right) = 12x\left(\frac{7}{12}\right) \text{ use distribute property left side}$$

$12x\left(\frac{2}{3}\right) + 12x\left(\frac{3}{4x}\right) = 12x\left(\frac{7}{12}\right)$ now carry out multiplication canceling out denominators!!

$4x(2) + 3(3) = x(7)$ carry out the multiplication both sides

$8x + 9 = 7x$ Subtract $8x$ both sides

$\underline{-8x \quad -8x}$

$9 = -\ x$ or $x = -\ 9$

Practice using multiplying by Common Denominator

Solve for variable

a) $\frac{p}{8} + \frac{p}{4} = 100$

b) $\frac{w}{5} + \frac{2}{3} = \frac{w}{3}$

c) $\frac{x}{7} + \frac{4x}{5} = \frac{13}{35}$

d) $\frac{2}{5} + \frac{9}{y} = \frac{7}{6}$

e) $\frac{3}{d} + \frac{5}{6} = \frac{1}{6d}$

f) $\frac{p+9}{3} = \frac{p-4}{8} + \frac{1}{2}$

Answers: a) 800/3; b) 5; c) 13/33; d) 270/23; e) – 17/5; f) – 72/5

2-5 Assignments, Use any Method You Prefer

1. $\frac{r}{7} + \frac{r}{21} = 336$

2. $\frac{w}{9} + \frac{w}{81} = 20$

3. $\frac{c}{24} + \frac{5}{12} = \frac{c}{2}$

4. $\frac{k}{15} + \frac{9}{30} = \frac{7k}{90}$

5. $\frac{p}{17} + \frac{6p}{51} = \frac{1}{3}$

6. $\frac{4}{5} + \frac{22}{t} = \frac{5}{6}$

7. $\frac{30}{n} + \frac{25}{11} = \frac{9}{22n}$

8. $\frac{h+5}{4} = \frac{h-4}{12} + \frac{1}{6}$

9. $\frac{5}{p} + \frac{7}{45} = \frac{7}{15}$

10. $\frac{13g}{14} + \frac{g}{7} = \frac{11}{35}$

11. $\frac{a+8}{11} + \frac{1}{4} = \frac{a+49}{44}$

12. $\frac{b+9}{7} = \frac{b-4}{8} + \frac{1}{2}$

13. $\frac{15}{x} + \frac{31}{3x} = \frac{7}{15}$

2-6 FORMULAS

A FORMULA is an equation that uses multiple unknowns to describe a relationship of fact. For our purposes, a formula will be any equation with more than one unknown (or variable).

formulas to find answers. So we need to know how to use formulas to our best advantage!

Which of these five equations are formulas given our definition??

1. $2x + 3y = 12$
2. $V = \frac{QR}{3}$
3. $2K - (5 + K) = 121$

4. $3x^2 + 2x + 1 = 100$
5. $I = Prt$

Equations 1), 2), and 5) are formulas.

Equation 1) has two unknowns, x and y.

Equation 2) has three unknowns, V, Q, and R.

Equation 5) has four unknowns, I, P, r, and t.

Equations 3) and 4) each has one unknown, note: Equation 3) has K as an unknown, even though it shows up twice. It can be reduced, when distributing and combining like terms, to $K - 5 = 121$. So there is only one value to find for the unknown K. Equation 4) has one unknown, x, even though it appears twice, and one time as a square. There is still only one value to find for the x. Therefore, equations 3) and 4) are not formulas.

The First use of formulas is to substitute known values in all but one of the unknown values, then use the algebra steps to solve for the remaining variable.

For example, the formula $V = IR$ is used by electricians, where

- V = Volts (for voltage),
- I = Amps of current through a circuit, and
- R = Ohms of resistance in the circuit.

What is the Volts given .23 Amps and 400 Ohms,

Lets make $I = .23$ and $R = 400$, and substitute into $V = IR$

So $V = IR = (.23)\,(400) = 92$. So the Volts is 92 Volts

What If there are 100 Volts, and .67 Amps, and we want to find the Ohms of resistance?

we make $V = 100$, $I = .67$ then substitute into $V = IR$ as

so $V = IR$ becomes

$100 = .67\,R.$ We need to then divide both sides by .67

$\div .67 \div .67$

$149.25 = R,$ so there is 149.25 Ohms of resistance.

So the strategy is solve in formulas is

a. substitute for all but one of the unknowns (variables),
b. use reduction and equality properties in correct order of operation so that the unknown (variable) is isolated on one side of the equation.

Example: A simple one: $y = 3x + 7$ is a linear equation, where (x, y) represents points on the line when graphed,

Lets Solve for x, given $y = 13$ Substitute 13 for y in $y = 3x + 7$ we get

$13 = 3x + 7$ now subtract 7 both side

$\underline{-7 \qquad -7}$

$6 = 3x,$ divide both sides by 3

$\underline{\div 3 \quad \div 3}$

$2 = x$ So $x = 2$ when $y = 13$ in $y = 3x + 7$

Example:

In business, the formula $I = Prt$ is an important one. It is called the SIMPLE INTEREST FORMULA, where I = Interest accumulated, P = money borrowed (or principal) r = interest rate, t = time of investment.

Find I if $P = 2500$, $r = .06$, and $t = \frac{1}{12}$. We substitute these values into the formula as

$$I = Prt = 2500\,(.06)\left(\frac{1}{12}\right) = \frac{2500\,(.06)\,(1)}{12} = 12.50$$

Find r if $I = 45$, $P = 4600$ and $t = .25$. We substitute these values into the formula $I = Prt$ as

$45 = 4600\,(r)\,.25$ or $45 = 1150\,r$, divide both sides by 1150

$\underline{\div 1150 \div 1150}$

$.012 = r$

A formula for the relationship between Celsius and Fahrenheit temperature degrees is $C = \frac{5}{9}(F - 32)$ with C the degree Celcius and F the degree Fahrenheit

If $F = 95$ degrees, what is Celsius? Our formula becomes

$C = \frac{5}{9}(95 - 32) = \frac{5}{9}(F - 32) = \frac{5}{9}(63) = 35$ so 95 degree Fahrenheit = 35 degree Celsius

If $C = -20$, what is Fahrenheit? Our formula becomes

$-20 = \frac{5}{9}(F - 32)$ use distributive property right side

$-20 = \frac{5}{9}F - \frac{5}{9}(32)$ or

$-20 = \frac{5}{9}F - 17\frac{7}{9}$ add $17\frac{7}{9}$ both sides

$\underline{+17\frac{7}{9} \quad + \quad 17\frac{7}{9}}$

$-2\frac{2}{9} = \frac{5}{9}F$ multiply both sides by $\frac{9}{5}$

$$\frac{9}{5}\left(-2\frac{2}{9}\right)=\frac{9}{5}\left(\frac{5}{9}F\right) \text{ or } \frac{9}{5}\left(-\frac{20}{9}\right)=\frac{9}{5}\left(\frac{5}{9}F\right) \text{ carry out the multiplication}$$

$-4 = F$ so –20 degrees Celsius is –4 degree Fahrenheit

Practice

a) For the formula $C = 2\,(3.14r)$ find C if $r = 2.3$
Find r if $C = 200$

b) For $I = Prt$ find I if $P = 3000$, $r = .03$, $t = 20$
Find t if $P = 11000$, $r = .09$, $I = 4620$

c) For $B = \dfrac{A}{(1+rt)^2}$ Find B if $A = 1000$, $r = .02$ and $t = 12$
Find A if $B = 1234$, $r = .065$ and $t = 365$

d) For $P = \dfrac{y}{2q} + 1$ Find P if $y = 88$, $q = 22$
Find q if $P = 10$, $y = 18$

Answers: a) C = 14.444; r = 31.847... ; b) I = 1800; t = 4.666.. ; c) B = 650.364...; A = 754375.821... ; d) P = 3; q = 1

2-6 Assignments

Solve these formulas

1. for $3xy + p = 12$

a) find y when $x = 8$ and $p = 2$ ____________

b) find p when $x = 5$ and $y = -8$ ____________

2. for $2.3r - y + 6 = 21$

a) find r when $y = 7$ ____________

b) find y when $r = 4.23$ ____________

3. for $\dfrac{5+2x}{3y} = 4$

a) find x when $y = 2$ ____________

b) find y when $x = -17$ ____________

4. for $2q - 8p + t^2 = 2p$

a) find q when $p = 3$ and $t = 4$ ____________

b) find p when $q = .5$ and $t = 7$ ____________

5. for $2x + 2y = A$

a) find y when $x = 5$ and $A = 7.8$ ____________

b) find A when $x = -3$ and $y = 3$ ____________

6. for $I = prt$

a) find r when $I = 600$, $p = 2340$ and $t = 3$ ____________

b) find p when $I = 212$, $r = .04$, and $t = 12$ ____________

7. for $A = 6.28\,r^2 + 6.28\,rh$

a) find h when $A = 600$ and $r = 4$ ____________

b) find A when $r = 3$ and $h = 6.5$ ____________

8. for $\frac{g-h}{k-m} = 45$

a) find h when $g = 4$, $k = 12$, and $m = 7$ ___________

b) find k when $g = 110$, $h = 20$, and $m = 31$ ___________

9. for $40 - 3y + x - 8w = 34 + 6y - 5x + w$

a) find w when $y = 11$ and $x = 2$ ___________

b) find y when $x = 20$ and $w = 10$ ___________

c) find x if $y = 2$ and $w = 1$ ___________

10. for $V = \frac{1 + rp}{\left(1 - \frac{r}{t}\right)^2}$

a) find V for $r = .06$, $p = 100$ and $t = 12$ ___________

b) find p for $V= 2300$, $r = .8$, and $t = 2$ ___________

2-7 LITERAL EQUATIONS

Sometimes you have a formula, and you need to solve for some variable other than the "standard" one. For instance, the formula for the perimeter P of a square with sides of length s is $P = 4s$. You might need to solve this equation for s, so you can plug in a perimeter and figure out the side length.

This process of solving a formula for a given variable is called "solving literal equations." One of the dictionary definitions of "literal" is "related to or being comprised of letters," and variables are sometimes referred to as literals. So "solving literal equations" seems to be another way of saying "taking an equation with lots of letters, and solving for **one letter in particular**."

At first glance, these exercises appear to be much worse than your usual solving exercises, but they really aren't that bad. You pretty much do what you've done all along for solving linear equations and other sorts of equation; the only substantial difference is that, due to all the variables, you won't be able to simplify your answers as much as you're used to. Here's an example of how "solving literal equations" works.

DISTANCE, SPEED, AND TIME

A classis science formula is $d = r \cdot t$ where d = distance traveled, r = speed or rate of travel, t = time

Suppose the rate traveled is $\frac{65 \text{ mi.}}{\text{hr}}$ and the time traveled is 10 hr. We let $r = \frac{65 \text{ mi.}}{\text{hr}}$ and $t = 10$ hr.

So $d = r\,t$ becomes $d = \frac{65 \text{ mi.}}{\text{hr}}(10 \text{ hr}) = 65(10) \text{ mi.} = 650 \text{ mi.}$

What if we need a formula for the time if we know the distance traveled and the speed.

So we want a formula for t "in terms of" d and r.

So we start with $d = r\,t$, to isolate t on one side, we divide both sides by r

$$\div r \quad \div r$$

$$\frac{d}{r} = t$$

So now we take the distance traveled and divide by the speed to get the time of travel

Example, the distance, d, traveled is 50 mi., and the speed, r is 5 miph.

$$\frac{d}{r} = t \text{ becomes } \frac{50 \text{ mi.}}{\left(\frac{5 \text{ mi.}}{1 \text{ hr}}\right)} = 10 \text{ hr} = t.$$

Now, figure a formula for r in terms of d and t . . . .

Another example:

Solve $4x + 5y = 100$ for y

We want to isolate y on one side, so

$$4x + 5y = 100 \text{ subtract } 4x \text{ both sides}$$

$$\underline{-4x \qquad -4x}$$

$$5y = 100 - 4x \text{ divide both sides by } 5$$

$$\underline{\div 5 \qquad \div 5}$$

$$y = \frac{100}{5} - \frac{4x}{5} \text{ simplifies to } y = 20 - \frac{4x}{5}$$

Geometry example:

The formula for area of a rectangle is $A = bh$, with A = area, b = base measure, and h = height measure.

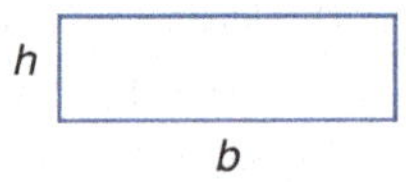

What is the formula for the base, b, in terms of the area, A, and the height, h?

In other words, Let's solve $A = bh$ for b in terms of A and h

$$A = bh \qquad \text{divide both sides by } h$$

$$\frac{A}{h} = \frac{bh}{h} \qquad \text{Cancel out the } h \text{ on the right side}$$

$$\frac{A}{h} = b,$$

What is the formula for the height in terms of the area and base?

PERIMETER OF A RECTANGLE

- Perimeter of a rectangle is the formula $P = 2l + 2w$ with P = perimeter, l = length, and w = width.

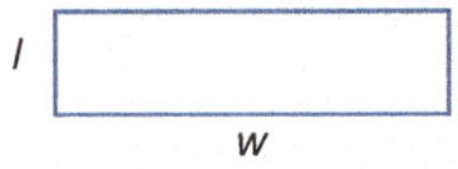

Let's solve $P = 2l + 2w$ for w in terms of P and l.

$$P = 2l + 2w \qquad \text{subtract } 2l \text{ both sides}$$

$$P - 2l = 2l + 2w - 2l$$

$$P - 2l = 2w \qquad \text{divide by 2 both sides}$$

$$\frac{P-2l}{2} = \frac{2w}{2}$$

$$\frac{P-2l}{2} = w$$

What is l in terms of P and w?

Solve $Q = \frac{c+d}{2}$ for d in terms of Q and d

$$Q = \frac{c+d}{2} \quad \text{multiply both sides by 2}$$

$$(2)\,Q = (2)\,\frac{c+d}{2}$$

$$2Q = c + d \quad \text{subtract } c \text{ both sides}$$

$$2Q - c = d$$

What is c in terms of Q and d?

Solve $12d + 13h - 4(d + 3) - 100 = 32d - 45h + 2r - 90$

For h in terms or d and r

Simplify left side, distributive property

$$12d + 13h - 4d - 12 - 100 = 32d - 45h + 2r - 90$$

Combine like terms left side

$$8d + 13h - 112 = 32d - 45h + 2r - 90$$

Use properties of equality to isolate h one side.

$$8d + 13h - 112 = 32d - 45h + 2r - 90$$

$$\underline{-8d \qquad\qquad\qquad\qquad -8d}$$

$$13h - 112 = 24d - 45h + 2r - 90$$

$$\underline{+112 \qquad\qquad\qquad\qquad +112}$$

$$13h = 24d - 45h + 2r + 22$$

$$\underline{+45h \qquad\qquad +45h}$$

$$58h = 24d + 2r + 22$$

$$\underline{\div 58 \qquad \div 58}$$

$$h = (24d + 2r + 22) \div 58 = \frac{24d + 2r + 22}{58}$$

$$h = \frac{12d + r + 11}{29}$$

Practice

a) Solve $T - W = 11$ for W in terms of T

For T in terms of W

b) Solve $12x = 4y + 30$ for y in terms of x

c) Solve $2R + 8B + 3C = 54 + 3R$ for B in terms of R and C

d) Solve $F = \frac{ma}{q}$ for a in terms of F, m, and q

For q in terms of F, m and a

Answers: a) T = 11 + W; b) y = (12x – 30) / 4 or y = (6x – 15) / 2; c) B = (54 + R - 3C)/8; d) a = Fq/m; q = ma/F

2-7 Assignments

Solve these formulas for the variable indicated in terms of the other variables.

1. Solve $y = 3x + 2$ for x _______________
2. Solve $r - x + 2y = 1$ for y _______________
3. Solve $A = .5bh$ for b _______________
4. Solve $6t + ry = \frac{n}{2}$ for y _______________
5. Solve $48t + 12k = 6(p - 2)$ for p _______________
6. Solve $2(x - h) - t = 3t$ for x _______________
7. Solve $\frac{1}{8}x + 4w = 3y$ for w _______________
8. Solve $V = 3.14r^2h$ for h _______________
9. Solve $.5\,h(k + 2t) = x$ for t _______________
10. Solve $y - 2a - 2b = -x + a + b$ for x _______________
11. Solve $\frac{3d + 7b - 9k}{2g + m} = 7$ for m _______________
12. Solve $\frac{a}{4c} + \frac{b}{2c} = 12$ for c _______________

2–8 PERCENTS

Knowing how to calculate percent's is an important skill to have. Here is a quick primer to help.

WHAT IS A PERCENTAGE?

Percent means "for every 100" or "out of 100." When we have a fraction with 100 in the denominator, the numerator is the percent of 100. The (%) symbol as a quick way to write a fraction with a denominator of 100.

For example, the fraction $\frac{13}{100}$ means 13 out of 100 and can be written as 13%. And $\frac{25}{100}$ is 25 out of 100 and can be written as 25%.

For example, instead of saying "it rained 37 days out of every 100 days," as 37 out of 100 is 37/100 or 37% we say "it rained 37% of the time."

Practice:

Write these fractions using the % notation

a) $\frac{45}{100}$ ________ **b)** $\frac{99}{100}$ ________

c) $\frac{1}{100}$ ________ **d)** $\frac{234}{100}$ ________

Answers:

a) 45%

b) 99%

c) 1%

d) 234%

Other Fractions can be changed into percent's. We can change a fraction into an equivalent fraction with a denominator of 100.

For example, $\frac{26}{200} = \frac{13}{100}$ so $\frac{26}{200}$ is 26 out of 200 which is 13%

Example: $\frac{97}{400} = \frac{24.25}{100}$, so $\frac{97}{400}$ is 97 out of 400 which is 24.25%

Practice:

Write these fractions as equivalent percent fraction and using the % notation

a) $a)\frac{8}{50}$ ________ **b)** $\frac{145}{300}$ ________

c) $\frac{1}{200}$ ________ **d)** $\frac{334}{800}$ ________

Answers:

a) 16/100, 16%

b) 48.33../100; 48.33..%

c) .5/100; .5 %

d) 41.74/100; 41.75%

Another way to figure the percent of a fraction is to simply divide the fraction into a decimal, then move the decimal two places to the right.

For example $\frac{86}{127} = 86 \div 127 \approx .677$ Moving the decimal two place to the right, we get 67.7 %

Example: $\frac{2,212}{768} = 2,212 \div 786 \approx 2.878$ Moving the decimal two place to the right, we get 287.8%.

Practice:

Write these fractions as decimal to nearest thousandths and as percent's

a) $\frac{11}{70}$ ________

b) $\frac{545}{1,200}$ ________

c) $\frac{8}{19}$ ________

d) $\frac{8,341}{701}$ ________

Answers:

a) .157; 15.7%

b) .454; 45.4%

c) .421; 42.1%

d) 11.899; 1189.9%

PERCENTS as PARTS OF WHOLE

Example, Harry took 45 bottles of shampoo out of the inventory of 350 bottles. What percent of the inventory did Harry remove?

If we think of the 350 bottle inventory as the WHOLE, the 45 bottles taken out as the PART, then the PERCENT is the PART Divided by the WHOLE

$$\text{The Percent} = \frac{PART}{WHOLE} = \frac{45\ Bottles}{350\ Bottles} \approx .129 \text{ or } 12.9\%$$

So, given P = Part, W = Whole, and Percent = R,

$$R = \frac{P}{W} \text{ or the Percent = Part divided by the whole}$$

It also follows that

$$P = R \cdot W \text{ or the Part = the Percent times the whole}$$

And $W = \frac{P}{R}$ or the Whole = the Part divided by the percent

Example: 75% of the 120 Beauty stores in the ALLBeauty Corporation show profitable growth in 2018. What was the number of stores that showed growth?

We want the part, so Use $P = R \cdot W$ with $R = 75\% = .75$ and $W = 120$. So

$$P = .75 \cdot 120 = 90 \text{ stores showing growth}$$

Example: There were 270 students in the Cosmetology program and that was 13.5% of the students at the University. How many students were at the University?

We want the whole, so use $W = \frac{P}{R}$ with P = 270 and $r = 13.5\% = .135$

$$W = \frac{270}{.135} = 2{,}000 \text{ students at the university}$$

Example: 23 out of 45 hair colorings were blond. What percent were blond?

We want the percent, so use $R = \frac{P}{W}$ with $P = 23$ and $W = 45$. And $R = \frac{23}{45} = .511$ so 51.1% were blond.

Practice

a) What is 18% of 50? ___________

b) 45 is 300% of what? _________

c) 59 of 3,000 is what %? ___________

d) 35% of 19 is what? ________

e) 42.5% of what value is 22? _____

f) What part of 235 is 11.2%? _______

g) 40 volunteers work on Mr. Biggs Campaign in Toledo. This is 24% of his whole campaign staff. How many were on staff? ______

h) There are 45 carnations out of 350 flowers in the bouquet at the banquet. What percent of flowers were carnations? _____

i) 23.5% of the 6,000 dogs in Houndsville are poodles? How many are poodles? ____

Answers

a) .18 × 50 = 9

b) 45/3.00 = 15

c) 59/3,000 = .020 so 2%

d) .35 × 19 = 6.65

e) 22/.425 = 51.76

f) $.112 \times 235 = 26.32$

g) 40/.24 = 167 staff members

h) 45/350 = .129 or 12.9% of the flowers

i) $.235 \times 6{,}000 = 1{,}410$ poodles

2-8 Assignments

1. Write $\frac{5}{8}$ as a percent. ____________
2. 816 is what percent of 6,800? ____________________
3. 140% of what is 7,840? ________________
4. Write 1,652/826 as a percent. ___________
5. If you had 67.5% of 1245388, what do you have? ________
6. There are 2,340 students enrolled at Tacoma Creative Arts School. 120 of these students are in their final year of classes. What percent of the students are in the final year? _________
7. The Lakewood Dodgers Won 56 games in the 2016 Greater Tacoma Softball League. The team had an 84.8% winning percentage out of the total number of games played that year. How many games were played by the Dodgers in 2016? ________
8. A recent report for the start-up company, Super Pharmaceuticals Company, showed a 234.5% growth in revenue. What is that growth rate in a decimal? __________
9. Forty-two delegates to the state Constitution Party Convention were over age 66. If the total number of delegates attending was 560, what percentage of the delegates were over age 66? _________
10. A certain brand of lightbulb was projected to have a 4.5% defective rate of lightbulbs. If the number of defective lightbulbs was 234, how many total lightbulbs were produced? _____________
11. A fish and chips dinner and a soft drink cost a diner $13.50. If the diner left a 15% tip, how much was the tip? __________
12. Norma Jean received an 8.9% raise for her job. If she was earning $65,000 a year, what was the amount of the raise? _________
13. Billy Bob went to the tractor store and found a tractor listed at $35,000. To help Billy Bob make a buying decision, the salesperson said he can have it for 85% of the list price. Billy Bob said – "I'll buy it!" What price did Billy Bob pay for the tractor? _______________
14. On a new microwave, there was a 9% sales tax of the list price that came out to $16.50. What was the list price? _________

15. Write $\frac{3}{4}$ as a percent ___________________

16. a) Saddles "R" Us found that .235 of their saddle buying customers prefer the Roy Rogers brand of saddle. What is that in a percent? __________

b) If there are 1,000 customers in all, how many prefer the Roy Rogers brand? ___________

17. In the year 2000, there were two types of Peruvian Lizards in the world, the Blue Peruvian Lizard and the Yellow Peruvian Lizard. There were 560 known blue Peruvian Lizards and 1,245 known yellow Peruvian Lizards in 2000. What percent of all Peruvian Lizards were yellow in 2000? _____

18. Quality TV's decided to decrease the price of their Wide Screen TV's by 19% which came out to a $345 decrease. What was the original price of the TV's? __________

19. Five out of seven people prefer SuperDuper Conditioner. What percent prefer SuperDuper?

____

20. If 100% of the 400 condo units at the VeryPosh Condominiums complex were occupied, how many units were occupied? ______________

CHAPTER 3

Systems of Measurement and Unit Conversion

3-1 MEASUREMENT SYSTEMS, CUSTOMARY AND METRIC

MEASUREMENT SYSTEMS: There have been thousands of measuring systems used throughout history in different civilizations. Today, there has evolved two main standard forms of measuring: The Customary (also known as the old English) System and The Metric System.

When writing quantities of measuring units we use a DENOMINATE NUMBER. A denominate number consists of a number value and a unit of measure. Examples of denominate numbers are 5 inches, 10 meters, 3456 pounds, .08 milligrams, 450 days, and 12 degrees.

Let's look at these two main measuring systems.

ENGLISH/CUSTOMARY – ***USED PRIMARILY IN THE U.S. (and not many other places!!!!)***

Origins of Measurements of length

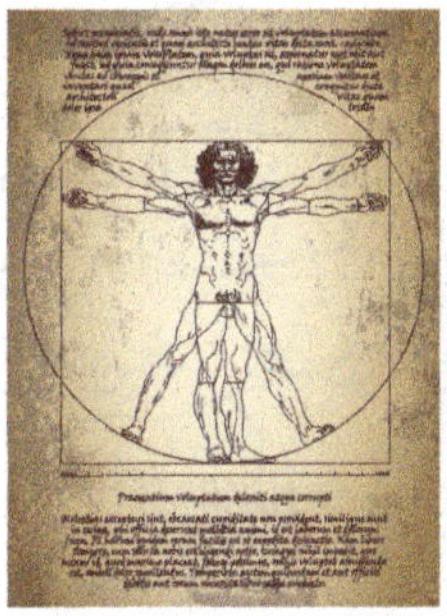

©Kwirry/Shutterstock.com

In ancient times, the **BODY** ruled when it came to measuring.

The length of a **foot**, the **width** of a finger, and the **distance of a step** were all accepted measurements.

©GCapture/Shutterstock.com

INCH: At first an inch was the width of a man's thumb.

©Swapan Photography/Shutterstock.com

In the 14th century, King Edward II of England ruled that 1 inch equaled 3 grains of barley placed end to end lengthwise.

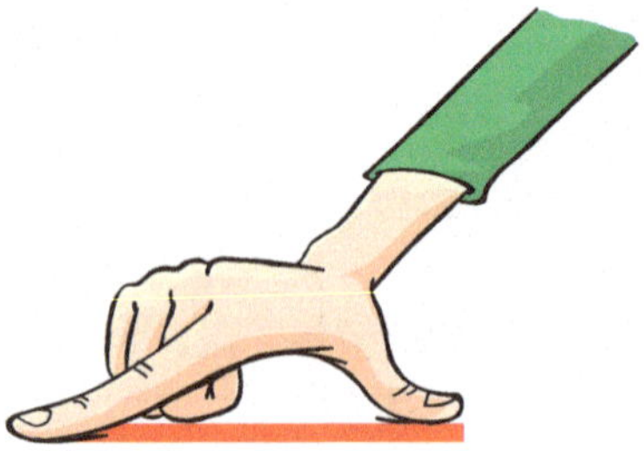

©Dusan Pavlic/Shutterstock.com

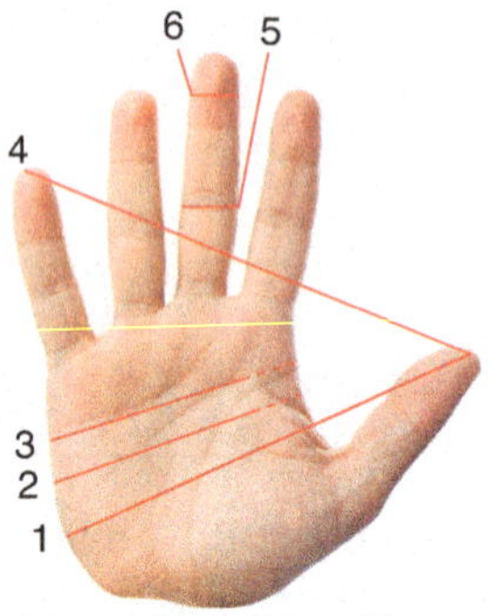

Adapted from ©Kues/Shutterstock.com; © Kendall Hunt Publishing

A hand was approximately 5 inches or 5 digits (fingers) across. Today, a hand is 4 inches and is used to measure horses (from the ground to the horse's withers, or shoulder).

SPAN: A span was the length of the hand stretched out, about 9 inches.

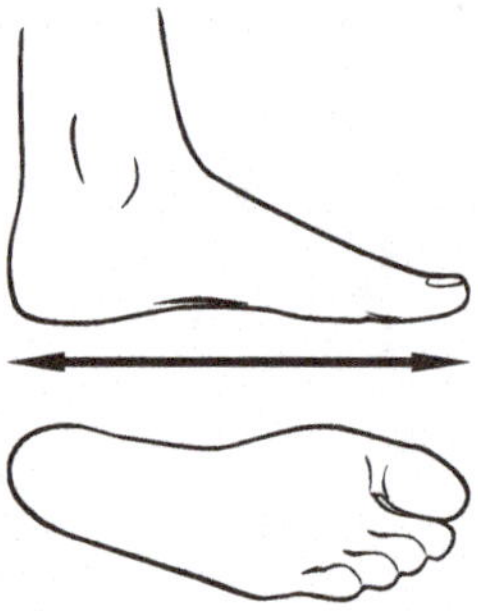

©BIELOUS NATALIIA/Shutterstock.com

FOOT: In ancient times, the foot was $11^1/_4$ inches. Today it is 12 inches, the length of the average man's foot.

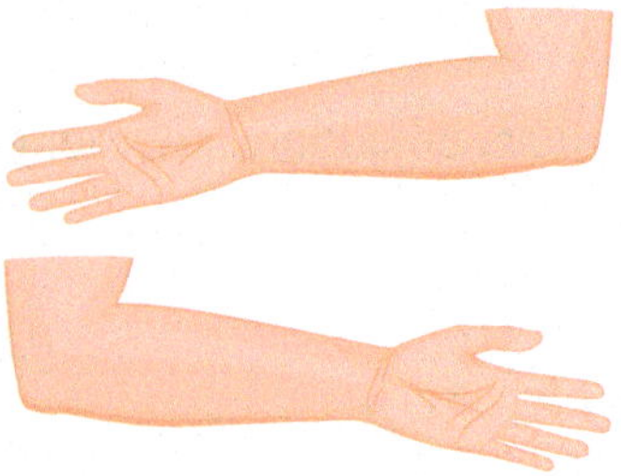

©Aksanaku/Shutterstock.com

YARD: A yard was originally the length of a man's belt or girdle, as it was called.

In the 12th century, King Henry I of England, (because her could, he was the king!!!), fixed the yard as the distance from his nose to the thumb of his out-stretched arm. Today it is 36 inches.

CUBIT: [use top image only and remove labels]

In ancient Egypt, a cubit was the distance from the elbow to the fingertips. Today a cubit is about 18 inches.

©Shendart/Shutterstock.com

[Only use the image on the left and put a line below his feet to measure between his feet]

PACE: The ancient Roman soldiers marched in paces, which were the length of a double step, . . about 5 feet; 1,000 paces was a mile. Today, a pace is the length of one step, $2^{1}/_{2}$ to 3 feet.

These measurements would be ok, except - everyone's body is different. So, depending on who was doing the measuring, measurements could vary. There was a need to create a **STANDARD SYSTEM OF MEASUREMENTS**:

And the English or Customary system we use today predominantly in the US has these basic standard units of measures,

- Inch. in. = 0.083 or 1/12 feet
- Foot, ft = 12 inches
- Yard, yd = 3 feet or 36 inches
- Mile, mi. = 5,280 feet or 1,760 yards

METRIC SYSTEM – used worldwide

The metric system was developed to utilize the base ten numbering system that was popular about the world.

METRIC LENGTH

The basic unit of measure for length in the metric system is the METER. A meter is about 3 feet (or a yard) in U.S. Measurement. We use m. to abbreviate meters.

The Meter was deemed an official measurement length by the French back in the 1700's. The officials in charge of measurements set it equal to be 10^{-7} of the distance from the equator to the pole on the meridian of the Earth though Paris.

That is $\frac{1}{10{,}000{,}000}$ of the distance you would walk in a straight line from the north pole to the equator. How they measured that distance in the 1700's - who knows, but they did it, then divided it by 10,000,000 and, voila, we have the meter length!!!!!!. Whew!

The metric system has prefix modifiers that apply to the basic meter unit, and create multiples of ten for new units. Here are the most common prefixes used. And **We find these prefixes apply not only to measure of length, but to other measures, such as volume and weight.**

Here are the most common prefix/multiples of ten used.

Prefix	Multiplier	Power of 10	Symbol
mega	1,000,000	10^6	M
kilo	1,000	10^3	k
meter	1		
centi	1/100	10^{-2}	c
milli	1/1,000	10^{-3}	m
micro	1/1,000,000	10^{-6}	mc or µ

Here is the breakdown for length measures:

- A megameter, Mm, is 1,000,000 (or 10^6) meters
- A kilometer, km, is 1000 (or 10^3) meters
- A meter, m, is the basic unit of length
- A centimeter, cm is 1/100 (or 10^{-2}) meter
- A millimeter mm, is 1/1000 (or 10^{-3}) meter
- A micrometer, mcm, is 1/1,000,000 (or 10^{-6}) meter

In reality, the kilometer (to measure really long distances), the meter, the centimeter (for small distances) and the millimeter (for even smaller distances) are the most common metric length measures.

For your information, Here is a more complete prefix, multiplier, power of ten and symbol chart.

Prefix	Symbol	Multiplication factor		
exa	E	10^{18}	=	1 000 000 000 000 000 000
peta	P	10^{15}	=	1 000 000 000 000 000
tera	T	10^{12}	=	1 000 000 000 000
giga	G	10^{9}	=	1 000 000 000
mega	M	10^{6}	=	1 000 000
kilo	K	10^{3}	=	1 000
hecto	h	10^{2}	=	100
deca	da	10^{1}	=	10
deci	d	10^{-1}	=	0.1
centi	c	10^{-2}	=	0.01
milli	m	10^{-3}	=	0.001
micro	µ	10^{-6}	=	0.000 001
nano	n	10^{-9}	=	0.000 000 001
Pico	p	10^{-12}	=	0.000 000 000 001
Femto	f	10^{-15}	=	0.000 000 000 000 001
Atto	a	10^{-18}	=	0.000 000 000 000 000 001

To give you an idea of how metric lengths compare to US measures, here is a sampling

FOR EXAMPLE:

1 kilometer (km) = 0.62 mile (mi)
1 kilometer (km) = 3280.8 feet (ft)
1 meter (m) = 3.28 feet (ft)
1 centimeter (cm) = 0.39 inch (in)
1 millimeter (mm) = 0.039 inch (in)
1 inch (in) = 2.54 centimeters (cm)
1 inch (in) = 25.4 millimeters (mm)
1 foot (ft) = 0.30 meter (m)
1 yard (yd) = 0.91 meter (m)
1 yard (yd) = 0.00091 kilometer (km)
1 mile (mi) = 1.61 kilometers (km)

THE FRACTION WOULD BE:

$$\frac{1 \text{ kilometer (km)}}{0.62 \text{ miles (mi)}} \quad \frac{\text{(numerator)}}{\text{(denominator)}}$$

VOLUME: To measure the space taken up inside solid figure, we use volume. Standard measures have developed in these units.

CUSTOMARY (US) VOLUME MEASURES

The common measures of volume in the U.S. system of measurements are:

- Teaspoons, tsp,
- Tablespoons, Tsp. = 3 teaspoons
- fluid ounces, fl. oz. = 2 tablespoons, 6 teaspoons
- cups, c. = 8 fluid ounces, 16 tablespoons
- pints, pt. = 2 cups, 16 fluid ounces
- quarts, qt. = 2 pints, 4 cups
- gallons, gal = 4 quarts, 8 pints, 16 cups

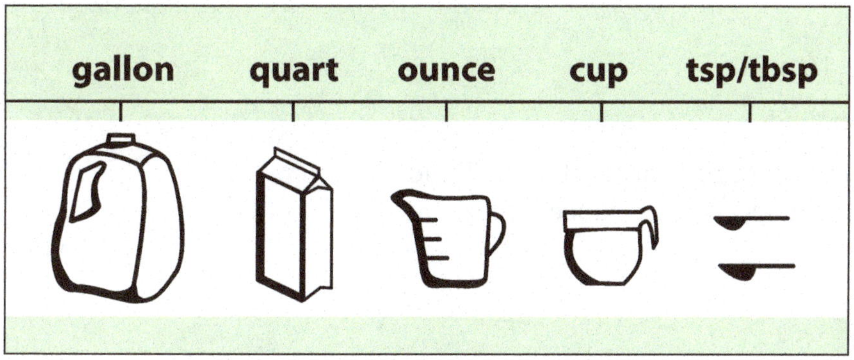

METRIC VOLUMES MEASURES

The basic volume measure is the LITER. One liter is slightly more than one Quart. We use the L. to abbreviate liters.

Applying the metric prefixes we get

Here is the breakdown for volume measures:

- A megaliter, ML, is 1,000,000 (or 10^6) liters
- A kiloliter, kL, is 1000 (or 10^3) liters
- A liter, L., is the basic unit of volume
- A centiliter, cL. is 1/100 (or 10^{-2}) liter
- A milliliter, mL, is 1/1000 (or 10^{-3}) liter
- A microliter, mcL, is 1/1,000,000 (or 10^{-6}) liter

Usually, the most common measures used for volume are the liter, L, and the milliliter, mL..

To give you a comparison of normal customary to metric volume measures:

1 Quart = .946 Liters;	1 liter = 1.057 Quarts
1 teaspoon = 4.93 milliliters;	1 milliliter = .203 teaspoons
1 Tablespoon = 14.787 milliliters;	1 milliliter = .068 tablespoons
1 cup = .237 liters;	1 liter = 4.227 cups
1 gallon = 3.786 liters;	1 liter = .264 gallons

MASS/WEIGHT - Mass is a measurement of the amount of matter something contains, while Weight is the measurement of the pull of gravity on an object. We often use the same units of measure for each, but weight can be different given the gravitational pull on the object (like the difference between being on the earth and being on the moon.)

The CUSTOMARY UNITS of WEIGHT MEASURE are

- Ounces, oz. a slice of bread is about an ounce in mass/weight.
- Pounds, lbs. = 16 ounces
- Tons, = 2000 pounds.

(Note, **fluid** ounces and ounces are different. Fluid ounces measures volume while ounces measure mass and weight.)

METRIC MASS/WEIGHT MEASURES

The basic mass/weight measure is the GRAM. A gram uses g. for abbreviation. One gram is about .036 of an ounce. Also there are about 28.35 grams in one ounce.

Here is the breakdown for volume measures:

- A megagram, Mg, is 1,000,000 (or 10^6) grams
- A kilogram, kg, is 1000 (or 10^3) grams
- A gram, g., is the basic unit of mass/weight
- A centigram, cg. is 1/100 (or 10^{-2}) grams
- A milligram, mg, is 1/1000 (or 10^{-3}) grams
- A microgram, mcg, is 1/1,000,000 (or 10^{-6}) grams

Usually, the most common measures used for mass are kilograms, grams, and milligrams,

To give you a comparison of normal customary to metric volume measures:

1 ounce = 28.35 grams;	1 gram = .036 ounce
1 Ounce = 28349.5231 Milligrams;	1 milligram = 0.0000352 ounce
1 kilograms = 2.205 pounds	

3-1 Assignments

For 1 – 8, True or false

1. Upper and lower case letters are Important to prefixes and abbreviations in metric measurement. __________
2. The Customary System of measurement is base 12. __________
3. The Metric System of Measurement is base 10. ____________
4. The abbreviation μL means microliters. __________
5. The abbreviation μ g means milligrams. ______________
6. There are about four liters in one gallon. _____________
7. There are about 450 grams in one pound. ______________
8. The Customary Measurement System is the one used universally throughout the world. ____

Write the best customary unit of measurement and the best metric unit of measure for measuring the following items

9. The length of your thumb
10. The length of your kitchen table
11. The fluid volume of a glass of water
12. The weight of a new born.
13. The fluid volume in a carton of milk.
14. The fluid volume of a gas tank of a car.
15. The weight of hard cover textbook.
16. The thickness of a quarter.
17. The weight of a marshmallow.
18. The length of a human foot.
19. The weight of a cell phone.
20. The weight of an automobile.

For the following measure, indicate the metric prefix that should be used for the basic unit of measure.

21. .01 grams = 1 ________ grams
22. 1000 Liters = 1 _______ Liters
23. .001 meters = 1 __________ meters
24. 10 grams = 1 ____ grams
25. 10^6 meters = 1 ______ meters
26. .000001 Liters = 1 _______ Liters
27. 100 grams = 1 _____ grams
28. 10^{-1} Liters = 1 _____ Liters

3-2 DIMENSIONS, DENOMINATE NUMBERS, AND OPERATIONS

A DIMENSION is a fundamental physical quantity in any system of measurement. There are seven fundamental dimensions in modern system of measurement. They are

1. LENGTH
2. MASS
3. TIME
4. TEMPERATURE
5. ELECTRICAL CURRENT
6. LUMINOSITY
7. AMOUNT OF SUBSTANCE

Units are measures used to quantify a dimension, and units associated with numbers quantify the units of measure. We do this when we create a DENOMINANT NUMBER. The number is coupled with the unit and the unit is coupled with the number.

So 12 ft., 5 lb., 98°, 45 hr, and so on, all describe some dimension, or identifies what IS BEING MEASURED OR COUNTER.

A denominate number is really an algebraic quantity.

We can think of

- The unit itself as sort of a variable.
- The number with the unit as the variables coefficient.

So "3 in." is like 3 in. or three times the inch measure.

43.5 mL is like 43.5 mL or 43.5 times the milliliter measure, and so on.

Thinking this way, we can apply the simplifying rules when adding (subtracting), multiplying (dividing), and using properties of equality with units of measures in equations.

> Here are some units of measure rules that follow the variable rule:
>
> - You can only add and subtract like units of measure.
> - You can multiply and divide both like and unlike units of measure.

ADDING AND SUBTRACTING

Just like simplifying $3x + 4x$ to $(3 + 4)x$ to $7x$, we can make 3 in. + 4 in. equal to (3 + 4) in. equal to 7 in.

However, $3x + 4y$ is simplified as we cannot add these two unlike term.

Same goes for 3 in. + 4 ft. Inches and feet are unlike units, so we cannot add them as is into one unit of measure.

How about 43 qt. – 23 qt. As we have like units, it becomes (43 – 23) qt. = 20 qt.

But we cannot subtract 43 qt. – 23 pt. as is. into one unit of measure

The only way to add or subtract unlike units of measure is to change (or convert) one of them in terms of the other, such as changing inches into feet.

Example: Find 5 in. + 4 ft. in feet.

As there are 12 in. in 1 ft., 5 in. is $\frac{5}{12}$ of a foot. So we can write

$$5 \text{ in.} + 4 \text{ ft.} = \frac{5}{12} \text{ ft.} + 4 \text{ ft.} = 4\frac{5}{12} \text{ ft.}$$

Now, find 5 in. + 4 ft. in inches.

As 4 ft. is 4(12) in. = 48 in., 5 in. + 4 ft. = 5 in. + 48 in. = 53 in.

Example: 14 gal. – 13 qt. in gallons.

As there are 4 qt. in 1 gal., 12 qt. is 3 gal., and 1 qt. is $\frac{1}{4}$ gal, so 13 qt. is $3 + \frac{1}{4}$ gal. $= 3\frac{1}{4}$ gal.

$$14 \text{ gal.} - 13 \text{ qt.} = 14 \text{ gal.} - 3\frac{1}{4} \text{ gal.} = 14 - (3 + \frac{1}{4}) \text{ gal.}$$

$$= (14 - 3 - \frac{1}{4}) \text{ gal.} = (11 - \frac{1}{4}) \text{ gal.} = 10\frac{3}{4} \text{ gal.}$$

Now, find 14 gal. – 13 qt. in quarts.

As 14 gal. is 14(4) qt. = 56 qt., then 14 gal. – 13 qt.

$$= 56 \text{ qt.} - 13 \text{ qt.} = 43 \text{ qt.}$$

DOUBLE DENOMINATE NUMBERS AND ADDITION/SUBTRACTION

What if we have double denominate numbers in addition or subtraction? One way is to change them to one unit of measure then add or subtract.

For example, 4 ft. 3 in. + 2 ft. 11 in.

We can change both into feet then add.

$$4 \text{ ft. } 3 \text{ in.} + 2 \text{ ft. } 11 \text{ in.} = 4\frac{3}{12} \text{ ft.} + 2\frac{11}{12} \text{ ft.} = (4 + \frac{3}{12} + 2 + \frac{11}{12}) \text{ ft.}$$

$$= (6 + \frac{14}{12}) \text{ ft.} = (6 + 1 + \frac{2}{12}) = 7\frac{2}{12} \text{ ft.} = 7\frac{1}{6} \text{ ft.}$$

Or change both into inches then add

$$4 \text{ ft. } 3 \text{ in.} + 2 \text{ ft. } 11 \text{ in.} = 48 \text{ in.} + 3 \text{ in.} + 24 \text{ in.} + 11 \text{ in.}$$
$$= (48 + 3 + 24 + 11) \text{ in.} = 86 \text{ in.}$$

Another way we can add double denominate number is to add the respective units of measure and create a final double denominate number.

4 ft. 3 in. + 2 ft. 11 in. = 4 ft. + 3 in. + 2 ft. + 11 in.

= 4 ft. + 2 ft. + 3 in. + 11 in. = 6 ft. + 14 in., and 14 in. = 1 ft. 2 in.,

Then 6 ft. + 14 in. = 6 ft. + 1 ft. + 2 in. = 7 ft. 2 in.

Practice

Add/subtract, give answers in both feet and inches

a) 12 ft. + 7 in. =

b) 14 ft. – 5 in. =

Add to quarts then to pints

c) 6 qt. + 7 pt. =

d) 8 qt. – 3 pt. =

Add/subtract and give as double denominate number

e) 6 ft. 6 in. – 3 ft. 8 in. =

f) 20 lb. 5 oz. + 5 lb. 12 oz. =

Answer

a) 12 7/12 ft.; 151 in.;

b) 13 7/12 ft.; 163 in.;

c) 9 1/2 qt.; 19 pt.;

d) 6 1/2 qt. ; 13 pt.;

e) 2 ft. 10 in.

f) 26 lb. 1 oz.

DIVIDING

$15y \div 3y$ can be simplified to $\frac{15y}{3y} = \frac{5}{1} = 5$

So is 15 yd. ÷ 3 yd. = $\frac{15\text{ yd.}}{3\text{ yd.}} = \frac{5}{1} = 5$. The units of measure cancel each other out, just like the variable y does in the previous expression.

Now $15x \div 3y = \frac{15x}{3y} = \frac{5x}{y}$

We cannot further simplify the $\frac{x}{y}$ as they are unlike

Same goes for 15 mi. ÷ 3 hr = $\frac{15\text{ mi.}}{3\text{ hr}} = \frac{5\text{ mi.}}{\text{hr}}$. This forms what is called a rate. A rate is a comparison of two different measure denominate numbers by division. We use rates all the time in industrial profession, such as miles/gallon, inches/hour, people/year, milligram per minute, and so on.

PRACTICE:

Find the rate for

a) 117 gal. ÷ 9 mi

b) For 678 in. ÷3 min

c) For 456 lb. ÷ 8 in.^{2}

Answer:

a) a) 13 gal./1 mi. ;

b) b) 226 in. / 1 min. ;

c) c) 57 lb./1 in^2

MULTIPLYING

When multiplying $6t\ 7p$, we multiply the numbers then multiply the variables to get $6t\ 7p = 6 \cdot 7 \cdot t \cdot p = 42\, t\, p$.

When multiplying units of measure, it is ok to multiply unlike units. Like variable factors, we multiply the numbers and we multiply the units of measure. For example,

$$6 \text{ ft.} \cdot 7 \text{ in.} = 6 \cdot 7 \cdot \text{ft. in.} = 42 \text{ “ft. in.”}$$

However, it may not always produce a *sensible* new unit of measure. We normally do not use “feet inches” as a unit of measure.

In this case, it makes sense to convert one of the denominate number measure to the other unit of measure.

For 6 ft.· 7 in., we can change the 6 ft. into in. As there are 12 in. in 1 ft., 6 ft. is (6·12) in. = 72 in.

So 6 ft. · 7 in. = 72 in. · 7 in. = 72 · 7 · in. · in. = 504 in.^{2} or 504 sq. in. A square inch is a recognized unit of measure for area.

Example:

$$4 \text{ ft. } 3 \text{ in.} \cdot 7 \text{ ft. } \ 7 \text{ in.}$$

Change to feet then multiply

$$4 \text{ ft. } 3 \text{ in.} = 4\frac{3}{12} \text{ ft.}, 7 \text{ ft. } 7 \text{ in.} = 7\frac{7}{12} \text{ ft.}$$

$$\text{So } 4 \text{ ft. } 3 \text{ in.} \cdot 7 \text{ ft. } 7 \text{ in.} = 4\frac{3}{12} \text{ ft.} \cdot 7\frac{7}{12} \text{ ft.} = 4\frac{3}{12} \cdot 7\frac{7}{12} \text{ ft} \cdot \text{ ft.}$$

$$= \frac{51}{12} \cdot \frac{91}{12} \text{ ft.}^2 = \frac{4641}{144} \text{ ft.}^2 = \frac{1547}{48} \text{ ft.}^2 = 32\frac{11}{48} \text{ ft.}^2$$

Or change to inches then multiply

$$4 \text{ ft. } 3 \text{ in.} = 48 \text{ in.} + 3 \text{ in.} = 51 \text{ in.}, 7 \text{ ft. } 7 \text{ in.} = 84 \text{ in.} + 7 \text{ in.} = 91 \text{ in.}$$

$$\text{So, } 4 \text{ ft. } 3 \text{ in.} \cdot 7 \text{ ft. } 7 \text{ in.} = 51 \text{ in.} \cdot 91 \text{ in.} = 51 \cdot 91 \cdot \text{in.} \cdot \text{in.} = 4641 \text{ in.}^2$$

NOTE: Conversion is not always needed, however, there are units of measures that use multiplication of individual units, such as light year (speed of light times years) or foot pounds (feet times pounds).

Practice

Multiply, give answers in both square feet and in square inches

a) 7 ft. · 18 in. =

b) 8 ft. 9 in. · 4 ft. 1 in. =

c) 2 ft. · 7 in. =

d) 5 ft. 10 in. (40 ft. 3 in.) =

Answer

a) 10 1/2 square feet; 1512 square inches;

b) 35 105/144 square feet; 5145 square inches;

c) 1 1/6 square feet; 168 square inches;

d) 234 57/72 square feet; 33810 square inches

3-2 Assignments

Convert these double denominate numbers to one denominate number

1. 12 feet 3 inches = ____________ feet
2. 12 feet 3 inches = ____________ inches
3. 2 miles 1320 feet = __________ feet
4. 2 miles 1320 feet = ___________ miles
5. 3 gallons 12 fl ounces = _________ gallons
6. 3 gal 12 fl oz = ____________ fl oz
7. 3 hr 38 min = _________ min
8. 3 T 750 lb = _______T

Perform the indicated addition or subtraction, write answer in simplified double denominate number

9. 9 ft 5 in + 12 ft 8 in ____________
10. 4 mi 3880 ft + 4 mi 2690 ft ______________
11. 13 gal 1 qt + 4 gal 3 qt ______________
12. 15 gal 2 qt – 5 gal 3 qt _____________
13. 3 hr 34 min + 6 hr 55 min ___________
14. 55 min 45 sec – 30 min 55 sec ____________

Perform the indicated division, Give as fractional unit rate

15. 1566 miles ÷ 6 hours ___________

16. 980 pounds ÷ 14 in^2 ____________

17. 672 miles ÷ 14 gallons _________

18. 5676 yard ÷ 6 hours ___________

Perform the multiplication, answer in unit of measure indicated

19. 8 feet times 11 inches = _________ sq feet

20. 4 feet times 8 yard = ___________ sq yards

21. 6 ft 3in times 3 ft 3 in = ____________ sq inches

22. 2 ft 9in times 4 ft 3 in = ___________ sq ft

23. 5 yds 1 ft times 6yd 2 ft = _________ sq ft

24. 9 yards 2 feet times 2 yards 1 foot = ___________ sq yards

3-3 CUSTOMARY SYSTEM UNIT CONVERSIONS

US CUSTOMARY SYSTEM OF MEASUREMENT

Here are the basic units of measure for the seven dimensions

Dimension	Basic unit of measure
LENGTH	INCH
MASS	POUND
TIME	SECOND
TEMPERATURE	DEGREE OF FAHRENHEIT
ELECTRICAL CURRENT	AMPERE
LUMINOSITY	CANDELA
AMOUNT OF SUBSTANCE	MOLE

We could have just stick to these basic units and figure everything in terms of inches, pounds, seconds, and so on. However, larger number of these units or small fraction of these units are just too cumbersome to write (like 20,000,000 in. or 1/64 of a pound) so to help ease this, other common units of measure evolve from multiples or fractional parts of these basic units.

For example, when we have 12 in., it is recognized that is equivalent to 1 ft.; when we have 2,000 lb., we say we have 1 tn., and so on.

As we have learned, we need to be able to convert between units of measure so we can apply the operations addition and subtraction, and in some cases multiplication, to like units of measure.

In order to be able to convert, we must first learn some basis conversion equations among measure in the US Customary System. Here are some …. These should be committed to memory!

US CUSTOMARY SYSYEM OF MEASUREMENT

LENGTH	1 ft. = 12 in.
	1 yd. = 3 ft. = 36 in.
	1 mi. = 5,280 ft. = 1,760 yd.
WEIGHT	1 lb. = 16 oz.
	1 tn. = 2,000 lb.
TIME	1 min = 60 sec
	1 hr = 60 min
	1 day = 24 hr
AREA	1 sq. ft. ($ft.^2$) = 144 sq. in. ($in.^2$)
	1 sq. yd. ($yd.^2$) = 9 sq. ft.
VOLUME	1 cup (c) = 8 fl. oz.
	1 pt. = 2 cups = 16 fl. oz.
	1 qt. = 2 pt. = 4 cups = 32 fl. oz.
	1 gal. = 4 qt. = 8 pt. = 16 cups = 128 fl. oz.
	1 cu. ft. ($ft.^3$) = 1728 cu. in. ($in.^3$)
	1 cu. yd. ($yd.^3$) = 27 cu. ft.
	1 gal. = 231 cu. in.

NOTE: For area, you can use sq. ft. for $ft.^2$ and for volume, you can use cu. ft. for $ft.^3$

To convert from one unit of measure to another we can

- Write the quantity you desire to convert as a fraction.
- Multiply by a fraction formed by the conversion equation (called a conversion identity) so the units not in the final answer cancel out.
- Simplify your final number value if needed.

HERE IS AN EXAMPLE: Convert 31 in. to feet.

First, write 31 in. as $\frac{31\text{ in.}}{1}$

Next, the conversion table shows that there are 12 in. in 1 ft. so 12 in. = 1 ft. As they are equivalent, we can write an identity fraction $\frac{1\text{ ft.}}{12\text{ in.}}$ and this is equal to 1. This makes this an identity conversion fraction.

We take our $\frac{31 \text{ in.}}{1}$ and multiply by $\frac{1 \text{ ft.}}{12 \text{ in.}}$ or

$$\frac{31 \text{ in.}}{1} \cdot \frac{1 \text{ ft.}}{12 \text{ in.}}$$ Now doing the multiplication

We get

$$\frac{31 \text{ in.}}{1} \cdot \frac{1 \text{ ft.}}{12 \text{ in.}} = \frac{31 \text{ in.} \cdot 1 \text{ ft.}}{1 \cdot 12 \text{ in.}} = \frac{31 \text{ in. ft.}}{12 \text{ in.}} = \frac{31}{12} \text{ ft.} = 2\frac{7}{12} \text{ ft.}$$

Note, the "inches" unit of measure cancel out in our final answer. This is why we made our identity conversion fraction $\frac{1 \text{ ft.}}{12 \text{ in.}}$ instead of $\frac{12 \text{ in.}}{1 \text{ ft.}}$.

EXAMPLE: Convert 3,260 min to hr

Write as fraction $\frac{3,260 \text{ min}}{1}$. As 60 min = 1 hr, and we want to cancel out minutes, our identity conversion fraction is $\frac{1 \text{ hr}}{60 \text{ min}}$

Multiply

$$\frac{3,260 \text{ min}}{1} \frac{1 \text{ hr}}{60 \text{ min}} = \frac{3,260 \text{ hr}}{60} = 54\frac{20}{60} \text{ hr} = 54\frac{1}{3} \text{ hr}$$

EXAMPLE: Convert 15 gal. to cubic feet

Write as fraction $\frac{15 \text{ gal.}}{1}$

We don't have a direct conversion from gallons to cubic feet. We do, however, have a conversion from gallons to cubic inches and then a conversion from cubic inches to cubic feet, so we will do a two-step conversion.

1 gal. = 231 cu. in. We want to cancel out gallons, so use the identity conversion fraction $\frac{231 \text{ in.}^3}{1 \text{ gal.}}$, and we multiply

$$\frac{15 \text{ gal.}}{1} \cdot \frac{231 \text{ in.}^3}{1 \text{ gal.}} = 3,465 \text{ cu. in.}$$

Now, write as fraction $\frac{3,465 \text{ in.}^3}{1}$

Since 1 cu. ft. = 1,728 cu. in. and we want to cancel out cubic inches we use the identity conversion fraction $\frac{1 \text{ ft.}^3}{1,728 \text{ in.}^3}$

Multiply $$\frac{3,465 \text{ in.}^3}{1} \cdot \frac{1 \text{ft.}^3}{1,728 \text{ in.}^3} = \frac{3,465}{1,728} \text{ cu. ft.} = 2\frac{9}{1,728} \text{ cu. ft.} = 2\frac{1}{192} \text{ cu. ft.}$$

We can also do this in one continuous equation where we line up our identities and multiply through

$$\frac{15 \text{ gal.}}{1} \cdot \frac{231 \text{ in.}^3}{1 \text{ gal.}} \cdot \frac{1 \text{ft.}^3}{1,728 \text{ in.}^3} = \frac{15 \cdot 231 \cdot 1 \text{ft.}^3}{1 \cdot 1 \cdot 1728} = \frac{3,465}{1,728} \text{ cu. ft.} = 2\frac{1}{192} \text{ cu. ft.}$$

PRACTICE:

Convert

a) 321 in. to yards

b) 14 mi. to feet

c) 600 oz. to pounds

d) 2,134 sec to minutes

e) 2,467 sq. in. to square yards

f) 463 cu. in. to quarts

Answers:

a) 8 11/12 yards;

b) 73920 feet;

c) 37 1/2 pounds;

d) 35 17/30 minutes;

e) 1 1171/1296 square yards;

f) 7 127/231 quarts

3-3 Assignments

Convert each quantity. Round the nearest hundredth of a unit when appropriate.

1. 192 inches = __________ feet
2. 11 pound = _______ ounces
3. 1 Ton = ________ ounces
4. 39 feet = _______ yards
5. 19.5 pints = ________ quarts
6. 4 miles = __________ feet
7. 2.3 miles = __________ yards
8. 656 ounces = _____ pounds
9. 9000 pounds = _____ Tons
10. 32 gallons = _____ cups
11. 428 gal = _____ qt
12. 9 ft^2 = _____ in^2
13. 1008 in^2 = _____ ft^2
14. 324 ft^3 = ______ in^3
15. 5 ft^3 = ______ qt.
16. 345611 oz. = ______ T.

17. 8 mi. = __________ in.

18. 324 yd^2 = __________ in^2

19. 1 yd^3 = ____________ gal.

20. 1 ft^3 = ______ qt.

21. 1 fl oz = _____ in^3

22. 1 in^3 = ______ fl oz.

23. 3450 min. = _______ hrs

24. 9201 hrs = ______ sec.

25. 45500 sec = ______ min

26. 10 qt/1 hr = ______ ft^3/1min

27. 5.5 yd^3/1min = ____ ft^3/1hr

28. 6.1 yd^3/hr _________ gal./min

29. 5670 in^3/sec = _______ pt./min

30. 3.45 T./hr = ______________ oz./sec

3-4 METRIC UNIT CONVERSIONS

Here are the basic unit of measure for METRIC in the seven dimensions

Dimension	Basic unit of measure
LENGTH	meter
MASS	gram
TIME	SECOND
TEMPERATURE	DEGREE OF Celsius
ELECTRICAL CURRENT	AMPERE
LUMINOSITY	CANDELA
AMOUNT OF SUBSTANCE	MOLE

In addition, for volume, the basic unit is the LITER.

Just like US customary, we could have just stick to these basic units and figure everything in terms of meters, grams, etc. However larger number of these units or small fraction of these units are just too cumbersome to write (like 20000000 meters or 1/1000 of a gram) Fortunately, Metric has a built in way of dealing with this, the POWERS OF TEN multipliers/prefix to represent larger or smaller units of the basic unit. So any conversion from metric to metric is either a multiplication or a division of a power of ten!!

Here are the most common prefix/powers of ten used in metric conversion.

prefix	multiplier	power of 10	symbol
mega	1,000,000	10^6	M
kilo	1,000	10^3	k
BASE UNIT	1		
centi	1/100	10^{-2}	c
milli	1/1,000	10^{-3}	m
micro	1/1,000,000	10^{-6}	mc or μ

Example:

Convert 50 grams to milligrams.

1 Milligram is 1/1000 of a gram. This also means there are 1000 milligrams in 1 gram.

We get $\frac{50\ grams}{1} \cdot \frac{1000\ milligrams}{1\ gram} = \frac{50000\ milligrams}{1} = 50000$ milligrams

Convert 8.4 centimeters to meters.

1 centimeter = 1/100 of a meter. This also means there are 100 centimeters in 1 meter.

We get $\frac{8.4\ centimeters}{1} \cdot \frac{1\ meter}{100\ centimeters} = \frac{8.4\ meters}{100} = .084$ meter

WE CAN USE THE DIFFERENCE IN POWERS TO FIGURE CORRECT IDENTITY FRACTION FOR CONVERSIONS

Convert 5 kiloliters to milliliters.

As kiloliters is 1000 liters or 10^3 liters and milliliters is 1/1000 liters or 10^{-3} *liters*, there is a difference 10^6 between kiloliters and milliliters. (3 to –3 is a difference of 6) So this means that there are 10^6 milliters in 1 kiloliters, or 1000000 milliliters = 1 kiloliter.

We now get $\frac{50\ kiloliters}{1} \cdot \frac{1000000\ milliliters}{1\ kiloliter} = \frac{5000000\ milliliters}{1} = 5000000$ milliliters.

Example, Convert 5768999 micrometers to centimeters.

As micrometers is 10^{-6} meters and centimeters is 10^{-2} meters, the difference of 10^4 between centimeters and micrometers $(-2 - -6 = 4)$. So this means there are 10^4 micrometers in 1 centimeter, or 10000 micrometers in 1 centimeters.

Therefore $\frac{5768999\ mcm}{1} \cdot \frac{1\ cm}{10000\ mcm} = \frac{5768999}{1000}$ cm = 576.8999 cm

Practice

a) Convert 689 Liters to kiloliters

b) Convert 754 milligrams to Kilograms

c) Convert 23 Megameters to Centimeters

d) Convert 56 microliters to milliliters

e) Convert 100 centigrams to Megagrams

Answers:

a) .689 kL;

b) .00754 kg

c) 2300000000 cm;

d) .056 mL;

e) .000001 Mg

3-4 Assignments

Convert these metric measurements

1. .0005 kilometers = ________ meters

2. 2.3 centimeters = ______ meters

3. 624 millimeters = ________meters

4. .045 Liters = _______ milliLiters

5. 4350 grams = ______ milligrams

6. 307 kilogram = _______ grams

7. 6.5 cm = _______ m

8. .8 kL = _______ L

9. 434 mm = _______ cm

10. 50 mcg = ______ mg

11. 5960 mL = _______ kL

12. .005 kg = __________ mg

13. 3.31 L = ______ mL

14. 9000 µg = _______ g

15. 45000000 cm = ________ Mm

16. 23456 L = _________ kL

17. 12300000 mcm = ________ km

18. .92 m = ________ mm

19. 123 Mg = ________ mg

20. 345000000 mL = ________ kL

3-5 CONVERSIONS BETWEEN CUSTOMARY AND METRIC SYSTEMS

We are often called upon to convert between the Customary and Metric families. These should be committed to memory!

US CUSTOMARY TO METRIC SYSTEM OF MEASUREMENT

LENGTH	I in. = 2.54 cm
	3.281 ft. = 1 m
	1 mi. = 1.609 km
WEIGHT	2.205 lb. = 1 kg

AREA	1 sq. in. = 6.452 cm^2
	10.765 sq. ft. = 1 m^2
VOLUME	1 cu. in. = 16.387 cm^3
	35.32 cu. ft. = 1 m^3
	1 fl. oz. = 29.57 mL
	1.057 qt. = 1 L
	1 gal. = 3.785 L
	1 gal. = 231 cu. in.

To convert from one unit of measure to another we can

- Write the quantity you desire to convert as a fraction.
- Multiply by a fraction formed by the conversion equation (called a conversion identity) so the units not in the final answer cancel out.
- Simplify your final number value if needed.

Convert 165 lb. to kilograms

As there are 2.205 lb. in 1 kg,

we get $\frac{165\text{ lb.}}{1} \cdot \frac{1\text{ kg}}{2.205\text{ lb.}} = \frac{165\text{ kg}}{2.205} = 74.83\text{ kg}$

Convert 37 cm to inches

As there are 2.54 cm in 1 in.

We get $\frac{37\text{ cm}}{1} \cdot \frac{1\text{ in.}}{2.54\text{ cm}} = \frac{37\text{ in.}}{2.54} = 14.57\text{ in.}$

Convert half a liter to fluid ounces

There are 1,000 mL in 1 L and there are 29.57 mL in 1 fl. oz., so

we get $\frac{1}{2} \cdot \frac{1\text{ L}}{1} \cdot \frac{1{,}000\text{ mL}}{1\text{ L}} \cdot \frac{1\text{ fl. oz.}}{29.57\text{ mL}} = \frac{1{,}000\text{ fl. oz.}}{2 \cdot 29.57} = 16.91\text{ fl. oz.}$

Convert 60 mph to kilometers per second

There are 1.609 km in 1 mi., 60 min in 1 hr and 60 sec in 1 min.

So we get $\frac{60\text{ mi.}}{1\text{ hr}} \cdot \frac{1.609\text{ km}}{1\text{ mi.}} \cdot \frac{1\text{ hr}}{60\text{ min}} \cdot \frac{1\text{ min}}{60\text{ sec}} = \frac{60 \cdot 1.609\text{ km}}{60 \cdot 60\text{ sec}} = \frac{.0268\text{ km}}{1\text{ sec}}$ or about .0268 km per sec.

Practice

a) 14 mi. to kilometers

b) 600 fl. oz. to milliliters

c) 2,134 kg to pounds

d) 2,467 sq. in. to square centimeters

e) 463 cu. in. to qt.

f) 2,340 mi. to meters

g) 2,345 mL to gallons

h) 13000 mg to pounds

i) $\frac{23 \text{ L}}{\text{day}}$ to $\frac{\text{fluid ounces}}{\text{hour}}$

Answers:

a) 22.536 km; b) 17742 mL; c) 4705.47 lb.

d) 15917.084 sq cm; e) 8.02 qt.; f) 3765060 m.;

g) .620 gal.; h) .029 lb; i) 32.41 fl. oz. / 1 hour

3-5 Assignments

Convert the following to the nearest hundredth when appropriate

1. 1 cm^2 = ______ m^2
2. 1 cm^3 = ______ m^3
3. 1 m^2 = ______ cm^2
4. 1 m^3 = ______ cm^3
5. 15 m^2 = ______ cm^2
6. 34200 cm^2 = ______ m^2
7. 56700000 cm^3 = ______ m^3
8. 11 m^3 = ______ cm^3
9. 480 mL = ______ fl oz
10. 242 lb = _______ kg
11. 6 ft = _______ m
12. 12 gal = ______ L
13. 20 mi = _______ km
14. 567 m = ______ ft

15. 5678 L = ______ qt

16. 367 kg = _____ lb

17. 12200 mL = ______ gal

18. 345 km = _____ ft

19. 20 gal = _____ mL

20. 18.4 ft^2 = ______ m^2

21. 45 m^2 = _____ ft^2

22. 43.2 cm^3 = _____ in^3

23. 9.2 L = _____ pt

24. 40 mi / 1 min = _____ km / 1 sec

25. 40 mi / 1 gal = ____ m / 1 L

26. 120 m / 1 hr = ____ ft / 1 sec

27. 15 lb/1 in^2 = ____ kg / 1 m^2

28. 782 kg/1 m^3 =_____ lb / 1 ft^3

29. 1 g / 1 cm^3 = ____ oz/ 1 in^3

30. 2.5 L / 1 day = ____ fl oz / 1 hr

31. 1600 cm^3/ 1 hr = _____ in^3/1 min

32. 7.5 km / 1 hr = _____ mi / 1 min

3-6 SOLVING SIMPLE MIXTURE PROBLEMS

It is useful as a Cosmetology or Esthetics professional to be able to handle mixing of two unknown quantities. Whether mixing two different percentage solutions into a new solution or mixing two different cost items into a new single cost item, being able to use the algebra to solve mixture problems makes the professional more competent and confidence in handling one's business needs.

Mixture problems have three amounts. Two of them are the amounts being mixed, and the third is the final mixture amount. Let's look at this with solution problems.

Solution Problems:

(% 1) (amount 1) + (% 2) (amount 2) = (final %) (total amount)

Example:

Wendell wants to mix a 15% cleaning solution with 23% cleaning solution to make 32 fluid ounces of 20% cleaning solution. How many fluid ounces of the 15% solution and how many fluid ounces of the 23% solution are needed to make the 32 fluid ounces of the 20% solution? Here are the steps:

- First, use a variable to represent the fluid ounces of one of the percentage solution amounts.
- Then use that variable to represent the other percentage solution amount.
- Put those two unknown variable representations into the equation and use simplifying expressions then properties of equality to solve for the variable.

Let A = the number of fluid ounces needed of the 15% cleaning solution. As there are 32 total fluid ounces in the final 20% solution, that means there are (32–A) fluid ounces of the 23% cleaning solution.

So in our formula ***(% 1) (amount 1) + (% 2) (amount 2) = (final %) (total amount)***

% 1 is 15%, amount 1 is A, % 2 is 23% , amount 2 is (32 − A), final % is 20%, and total amount is 32. Therefore, we get

15% of A plus 23% of (32 – A) equals 20% of 32 or

$.15A+.23(32-A)=.20(32)$ **Simplifying both sides**

$.15A+7.36-.23A=6.4$ **or**

$.15A-.23A+7.36=6.4$ **or**

$-.08A+7.36=6.4$ **Subtract 7.36 both sides**

$-7.36 \quad -7.36$

$-.08A=-.96$ **Divide both sides by –.08**

$\div -.08 \quad \div -.08$

$A=12$

So A = 12, and 32 − A = 32 = 12 = 20. And, that means there are 12 fluid ounces of the 15% cleaning solution and 20 fluid ounces of the 23% cleaning solution.

Example:

A chemist needs to mix an 18% fragrance solution with a 45% fragrance solution to obtain 12 liters of a 36% fragrance solution. How many liters of each of the fragrance solutions must be used?

Let F = number of liters of 18% fragrance solution and, as there are 12 liters in the mixed solution, let $(12 - F)$ = number of liters of the 45% fragrance solution.

% 1 is 18%, amount 1 is F, % 2 is 45%, amount 2 is (12−F), final % is 36%, and total amount is 12. Therefore, we get

18% of *F* plus 45% of (12 – *F*) equals 36% of 12 or

$.18F + .45(12 - F) = .36(12)$ **Simplifying both sides**

$.18F + 5.4 - .45F = 4.32$ ***or***

$.18F - .45F + 5.4 = 4.32$ ***or***

$-.27F + 5.4 = 4.32$ **Subtract 5.4 both sides**

$$\begin{array}{rcl} -5.4 & & -5.4 \\ \hline -.27F & = & -1.08 \end{array}$$ **(Divide both sides by −.27)**

$$\begin{array}{rcl} \div -.27 & & \div -.27 \\ \hline F & = & 4 \end{array}$$

So $F = 4$ and $12 - F = 12 - 4 = 8$. And, that means there are 4 liters of the 18% fragrance Solution and 8 liters of the 43% fragrance solution in the 12 liters of 36% fragrance solution.

Mixture Problems:

(cost 1) (amount 1) + (cost 2) (amount 2) = (final cost) (total amount)

Example:

Mindi sells potpourri at her beauty shop. She has in stock cinnamon scented potpourri and apple scented potpourri. Her cinnamon potpourri sells at $4.50 per ounce and her apple potpourri sells at $3.75 per ounce. She wants to create a mixture of the two potpourri scents and sell it as a new apple/cinnamon potpourri at $4.20 an ounce. If she makes 10-ounce packages of the apple/cinnamon mixture for sale, how many of ounce of the cinnamon and how many ounce of the apple potpourri would she use for a fair priced product?

Here are the steps

- First, use a variable to represent the ounces of one of the amounts in stock.
- Then use that variable to represent the other amount in stock.
- Put those two unknown variable representations into the equation and use simplifying expressions then properties of equality to solve for the Variable.

Let C = the ounces of Cinnamon potpourri needed. As there is 10 ounces in the mixture, then (10 – C) ounces is needed of the Apple potpourri.

So for our formula

(cost 1) (amount 1) + (cost 2) (amount 2) = (final cost) (total amount)

Cost 1 is $4.50; amount 1 is C, cost 2 is $3.75, amount 2 is (10 – C), final cost is $4.20, total amount is 10 therefore we get

$4.50 times C + $3.75 times (10 – C) equals $4.20 times 10 or

$\$4.50\ C + \$3.75\,(10 - C) = \$4.20\,(10)$ **Simplifying both sides**

$\$4.50C + \$37.50 - \$3.75C = \$42.00$ **or**

$\$4.50C - \$3.75C + \$37.50 = \$42.00$ **or**

$\$.75\ C + \$37.50 = \$42.$ **Subtract $37.50 both sides**

$-\$37.50 \quad -\$37.50$

$\$.75C = \$4.50$ **Divide both sides by $.75**

$\div\$.75 \quad \div\$.75$

$C = 6$

$C = 6$, $10 - C = 10 - 6 = 4$, so Mindi mixed 6 ounces of Cinnamon potpourri and 4 ounces of Apple potpourri for the new 10 ounce Cinnamon/Apple potpourri mixture package which sells for $42.00 a package.

Example:

Jen and Berry Ice Cream Shoppe want to mix Decadent Chocolate Ice Cream with Crazy Coconut Ice Cream to make a Choco-Coconut Ice cream. The Decadent Chocolate Ice Cream sells for $3.40 a pint and the Crazy coconut sells for $5.00 a pint. Jen and Berry want to sell a 4-pint container of Choco-Coconut Ice cream at a fair price of $4.50 a pint. How many pints of Decadent Chocolate Ice and how many pints of Crazy Coconut Ice Cream should be mixed for the Choco-Coconut Ice Cream 4 pint container.

Let D = number of pints of Decadent Chocolate Ice Cream. As there is 4-pints in the container, $(4 - D)$ will be the number of pints in the Crazy Coconut Ice Cream

And Cost 1 is \$3.40; amount 1 is D, cost 2 is \$5.00, amount 2 is $(4 - D)$, final cost is \$4.50, total amount is 4 therefore we get

\$3.40 times C plus \$5.00 times (4 – D) equals \$4.50 times 4 or

$\$3.40D + \$5.00(4 - D) \quad = \$4.50\,(4)$ Simplifying both sides

$\$3.40D + \$20.00 - \$5.00D = \$18.00$ or

$\$3.40D - \$5.00D + \$20.00 = \$18.00$ or

$-\$1.60\,D + \$20.00 = \$18.00$ Subtract \$20.00 from both sides

$-\$20.00 \qquad -\$20.00$

$-\$1.60D = \$-2.00$ Divide both sides by –\$1.60

$\div -\$1.60 \qquad \div -\$1.60$

$D = 1.25$

$D = 1.25,\ 4 - D = 4 - 1.25 = 3.75$ so Jen and Berry mixed 1.25 pints of Decadent Cholate Ice Cream with 3.75 pints of Crazy Coconut Ice Cream to make a 4 pint container of Choco-Coconut Ice cream, sold at 4 times \$4.50 = \$18.00 a container.

Here is one with a twist.

If 4 pounds of a candy that costs \$1.50 a pound is mixed with some pounds of another candy costing \$2.00 a pound, and the resulting candy mixture cost \$1.80 a pound, how many pounds are in the \$2.00 a pound candy and how many pounds total are there in the candy mixture.

Let M = the number of pounds of the \$2.00 a pound candy. That means, when combined with the 4 pounds of the \$1.50 a pound candy, there is $(M + 4)$ pounds of the mixture.

So, Cost 1 is \$1.50; amount 1 is 4, cost 2 is \$2.00; amount 2 is M, final cost is \$1.80, total amount is $(M + 4)$ therefore we get

$\$1.50(4) + \$2.00(M) = \$1.80(M + 4)$ Simplifying both sides

$\$6.00 + \$2.00M = \$1.80M + \$7.20$ Subtracting \$6.00 both sides

$-\$6.00 \qquad -\$6.00$

$\$2.00M = \$1.80M + \$1.20$ Subtracting \$1.80M both side

$-\$1.80M \qquad -\$1.80M$

$.20M = \$1.20$ Dividing both sides by .20

$\div .20 \qquad \div .20$

$M = 6$

$M = 6,\ M + 4 = 6 + 4 = 10$, so 6 pounds of the \$2.00 a pound candy mixed with 4 pounds of the \$1.50 a pound candy makes 10 pounds of the mixture. In addition, at \$1.80 pound, the 10 pounds cost \$1.80 times 10 pounds = \$18.00.

Practice

1. How many liters of a 90% alcohol solution and how many liters of 40% alcohol solution are mixed together to produce 15.5 liters of 50% alcohol solution?
2. In her store, Shelley sells bracelets contain silver and gold beads. Silver bead are $1.30 a bead and gold beads are $3.50 a bead. If a Bracelet sells for $43.50, contain 25 beads, and the beads make up the fair cost value of the bracelet, how many silver and how many gold beads are in a bracelet?
3. Geno started two investment plans in his investment portfolio. During the first year, The Big Spender plan earns interest of 8% and the Mammoth Value plan earns interest of 11%. In total, Geno invested $4,000, and the average interest earn for both plan came out to 9.5%. How much did Geno invest in the Big Spender plan and how much in the Mammoth Value plan?
4. Four bottles of Special Shine Shampoo at $8.50 a bottle was combined with some bottles of BestFresh Shampoo at $10.00 a bottle in a special SuperShampoo package sold at an average of $9.00 a bottle. To be fair, how many bottles of the Best Fresh shampoo and how many bottle totals in the SuperShampoo set were there? In addition, how much will the SuperShampoo set sell for?

Answers.

1. Let A = liters of the 90% solution. .90 A + .40(15.5 – A) = .50(15.5) or .90A – .40A + 6.2 = 7.75 or .50A = 1.55 or A = 3.1, 15.5 – 3.1 = 12.4, so 3.1 liters of 90% and 12.4 liters of 40%.
2. Let *A* = number of Silver Beads. $1.30A + $3.50(25–A) = $43.50 or $1.30A – $3.50A + $87.50 = $43.50 or –$2.20A + $87.50 = $43.50 or –$2.20A = –$44.00 or A = 20 And 25 – 20 = 5, so 20 Silver Beads and 5 gold beads
3. Let *B* = dollars invested in the Big Spender plan. .08 *B* + .11 ($4000 – *B*) = .095 ($4,000) or .08B + $440 – .11B = $380 or .08B – .11B + $440 = $380 or –.03B + $440 = $380 or –.03B = –$60 or $2,000 = B, and $4,000 – $2,000 = $2,000, So $2,000 of Big Spender Plan and $2,000 or Mammoth Value plan
4. Let *F* = number of bottle of BestFresh Shampoo, $8.50 (4) + $10.00(F) = $9.00 (F + 4) or $34.00 + $10.00F = $9.00F + $36.00 or $10.00F = $9.00F + $2.00 or $1.00F = $2.00 or F = 2. And 2 + 4 = 6, so 2 bottles of Best Fresh shampoo and 6 bottles total in the SuperShampoo set. The set will sell for 6 × $9.00 or $36.00

3–6 ASSIGNMENTS

1. A mixture solution is made up of 9% solution and 19% solution. If the final mixture is 50 milliliters and is now a 15% solution, how many milliliters of the 9% solution and how many milliliters of the 19% solution is mixed together? _______ _________

2. Kendell has a store where he mixed regular milk and chocolate milk into quart containers and sold it as "Kendall's Moo Dream" refreshment at $3.40 a quart. If Kendell sold the regular milk

separately at $2.80 a quart and the chocolate milk separately at $3.80 a quart, how much regular milk and how much chocolate milk does Kendell mix for the quart container mixture?

__________ ____________

3. FrannieBird's Specialty Salts made flavored salts. They had two garlic-flavored salt, the "Mildly Garlic "salt, with 6% granulated garlic, and the "That's So Garlic" salt with 16% granulated garlic. The company decided to make and sell a 40-ounce container with 10% granulated garlic and call it "Garlic Perfection" Salt. How much "Mildly Garlic" and how much "That's so Garlic" salt was needed to make up the "garlic Perfection "salt container? ________ ______________

4. Dr. Cy Intest makes up two chemical solutions with each containing mercury. Solution A contains 5% mercury and solution B contains 13% mercury. He needs a 15 fluid ounce container of solution that is 9% mercury. He mixes Solution A and Solution B in order to get the new mixture with 9% mercury. How much Solution A and how much solution B should Dr. Intest use for the 15 fluid ounce container?

____________________ ______________________

5. DebbiDelites Candy Shoppe sells red jellybeans for $2.50 a pound and green jelly beans for $3.25 a pound. The store decides as a Christmas Special to sell a mixture of the red and green jellybeans in a 3-pound bag and sell it at $2.80 a pound. How many pounds of red jellybeans and how many pounds of green jellybeans should go into the 3-pound bag of the mixture. ____________ ______________

6. Two salad dressings contain oil in different amounts. Salad Dressing A has 15% oil and Salad Dressing B has 5% oil. If we want to combine salad dressings A and B into another salad dressing with 11% oil, how much of Salad Dressings A and B would be needed to make 32 fluid ounces of the new salad dressing?

_______________ ______________________

7. Paulie sells fake fingernail sets at her beauty store. The Kool Nails set sells for $25.00 and the FancyNail set sells for $30.00. Paulie wants to offer a WonderNails package set mixing 3 Kool Nails sets and a number of FancyNail set and sell it for a fair price of $195 for the package. How many FancyNail sets are needed to complete the WonderNails package set? _____________

8. Delia wants to sell 4-pound packages of a mixture of peanuts and cashews in her store. If peanuts cost $1.70 per pound and cashews cost $2.50 per pound, how many pounds of peanuts and how many pounds of cashews are need to mix in order to charge a fair price of $2.30 a pound for the mixture package. ________ __________

CHAPTER 4

Time and Time Management

As a cosmetologist and esthetics professional, you need to be aware of time and how to manage it. For example, in a beauty shop, time is used to determine when to schedule the next appointment or how long to leave a permanent in place. And, you need to be able to figure the total hours you and/or your employees have worked.

4-1 CUSTOMARY AND MILITARY TIME

CUSTOMARY TIME SYSTEM

A lot of the world uses a CUSTOMARY time system (sometimes referred to Regular or Civilian time), which runs a day cycle in hours, minutes, and seconds, based on

- AM (ante-meridian)—where the day follows a start time of 12:00 AM (midnight), 1 through 11 hours AM to 12:00 PM (noon); and
- PM (post-meridian) system— where the day continues from 12:00 PM (noon), 1 through 11 hours PM back to 12:00 AM (midnight) , start of the next day.

MILITARY TIME SYSTEM

However, some businesses, like military and hospitals, use what is called the Military time system. Military time follows a day cycle where midnight starts of the day then each hour/minute of the day follows a 1 through 24 hour numbering system.

For example, the start of the day, 12:00 AM (midnight) in customary time, is 0000 in military time. Then 1:00 AM in customary time is 0100 (pronounced OH 100) in military time. 7:15 AM in customary time is 0715 (pronounced OH 715) in military time.

12:00 noon is 1200 (pronounced 12 hundred) in military time.

PM times in customary time is changed to military time by adding 12 to the hour number. So given 2:00 PM in customary time, then 12 + 2 = 14, so 2:00 PM is 1400 in military time. Given 5:35 PM in customary time, then 12 + 5 = 17, so 5:35 PM becomes 1735 in military time. Given 11:59 PM in customary time, then 11 + 12 = 23, so 11:59 PM becomes 2359 in military time.

Military	Customary	Military	Customary
0:00 =	12:00 AM	12:00 =	12:00 PM
1:00 =	1:00 AM	13:00 =	1:00 PM
2:00 =	2:00 AM	14:00 =	2:00 PM
3:00 =	3:00 AM	15:00 =	3:00 PM
4:00 =	4:00 AM	16:00 =	4:00 PM
5:00 =	5:00 AM	17:00 =	5:00 PM
6:00 =	6:00 AM	18:00 =	6:00 PM
7:00 =	7:00 AM	19:00 =	7:00 PM
8:00 =	8:00 AM	20:00 =	8:00 PM
9:00 =	9:00 AM	21:00 =	9:00 PM
10:00 =	10:00 AM	22:00 =	10:00 PM
11:00 =	11:00 AM	23:00 =	11:00 PM

Finally, upon reaching midnight 12:00 AM for the next day, Military time becomes 0000 again.

Here is a chart that gives the key Military to Customary conversion.

Practice:

Change Military time to Customary time (with AM or PM)

0112 ____________ 1013 ____________ 1523 ____________

0500 ____________ 1155 ____________ 1320 ____________

Change Customary time to Military time

8:30 AM ____________ 11:21 AM ____________ 5:20PM ____________

12:03 PM ____________ 1:21 AM ____________ 12 noon ____________

Answers: military to customary 1:12 AM , 10:13 AM, 3:23 PM, 5:00 AM, 11:55 AM , 1:20 PM

Customary to Military 0830, 1121, 1720, 1203, 0121, 1200

4-1 Assignments

For these times, write the time in the alternate time,
For example, given customary time 3:45 PM, write in Military time 1545

Given Military time 2233, write in Customary time 10:33 PM

1740 ____________ **6:50 PM** ____________ **1835** ____________

0040 ____________ **11:11 PM** ____________ **2030** ____________

1920 ____________ **10:10 PM** ____________ **1305** ____________

5:55 PM ____________ **1510** ____________ **8:15 AM** ____________

2310 ________________	0115 ________________	3:55PM ______________
7:05 PM ______________	0340 ________________	1745 ______________
1946 ________________	1819 _________________	1:55PM ______________
3:59 PM ________________	0009 _______________	8:20PM ______________
0235 ___________________	2:33 PM ______________	1855 ______________
0505 ____________________	10:40 AM _______________	0000 ______________
1:11 AM ________________	1:11 PM _______________	1200 ______________

4-2 ELAPSE TIME FROM CUSTOMARY TIMES

It is important the cosmetologist professional is able to calculate elapse time given a starting customary time and an ending customary time. We use a sort of subtraction operation in order to come up with the total elapse time.

- Turn 12 midnight to 12:59 AM and 1:00 PM to 11:59 PM customary times to Military time;
- Put the ending time on top of the starting time for a "subtraction" operation;
- Separate the minutes from the hours with a vertical line;
- Subtract the minutes. If the starting minutes is more than the ending minutes, you need to "borrow" 1-hour worth of minutes, or 60 minutes, from the ending hours, then subtract; and
- Subtract the hours.

For example, find the elapse time from 8:30 AM to 11:45 AM

11:|45

−8:|30 as we can subtract 30 from 45 and 8 from 11, we simply subtract.

3:|15 It is 3 hours 15 minutes elapse time.

Find the elapse time from 9:53 AM to 11: 04AM

11:|04

−9:|53 we can't subtract 53 from 4, so we borrow 1-hour, or 60 minutes, from 11 hour.

Which changes the end time from 11:04 to 10:64

To figure elapse time

10:	64	
−9:	53	Now we can subtract 53 from 64 and 9 from 10, so we do the subtractions.
1:	11	It is 1 hour 11 minutes elapse time

Find the elapse time from 10:22 AM to 2:33 PM

Turn 2:33 PM into 14:33 military time

14:	33	
−10:	22	as we can subtract 22 from 33 and 10 from 14, we do the subtraction
4:	11	It is 4 hours, 11 minutes elapse time

Find the elapse time from 9:46 AM to 12:30 PM

Turn 4:30 PM into 16:30 military time

16:	30	
−12:	46	We can't subtract 46 from 30, so borrow 1 hour (60 minutes) from 16
		Changing ending time from 16:30 to 15:90
15:	90	
−12:	46	subtract 46 from 90 and 12 from 15 to get
3:	44	3 hours 44 minutes elapse time

Find the elapse time from 5:50 PM to 10:27 PM

Turn 5:50 PM into 17:50 Military time, Turn 10:27 PM into 22:27 Military time

22:	27	
−17:	50	We can't subtract 50 from 27, so borrow one hour (60 minutes) from 22
		Changing the ending time from 22:57 to 21: 87
21:	87	
17:	50	We can subtract 50 from 87 and 17 from 21, so doing the subtractions
4:	37	It is elapse time of 4 hours 37 minutes

Find the elapse time from 12:22AM to 12:45PM

Turn 12:22 AM into 0:22 military time

12:	45	
−0:	22	we can subtract 22 from 45 and 0 from 12, so doing the subtracting
12:	23	It is 12 hours and 23 minutes elapse time

Practice

a) Find elapse time from 5:30 AM to 12:45 PM ________________

b) Find elapse time from 6:45 AM to 8:23 AM __________________

c) Find elapse time from 11:23 AM to 6:30 PM ________________

d) Find elapse time from 1:50 PM to 6: 25 PM ________________

e) Find Elapse time from 12:55 AM to 11:56 AM ______________

Answers:

a) 7 hours 15 minutes

b) 1 hours 38 minutes

c) 7 hours 7 minutes,

d) 4 hours 35 minutes

e) 11 hours 1 minute

4-2 Assignments

Find the elapse time in hours and minutes. SHOW ALL WORK

1. **9:18 AM to 3:15 PM** ___________
2. **10:00 AM to 5:48 PM** __________
3. **4:30 AM to 11:45 AM** __________
4. **8:19 AM to 9:02 AM** __________
5. **7:58 PM to 9:11 PM** __________
6. **9:00 AM to 9:57 PM** __________
7. **2:02 PM to 4:58 PM** __________
8. **12:43 PM to 5:06 PM** __________
9. **12:10 AM to 11:07 PM** ___________
10. **7:45 AM to 3:44 PM** ____________

In problems #11 —#15 find elapsed time for AM and PM and added together.

	AM		PM		
	IN	OUT	IN	OUT	
11.	**7:58**	**11:49**	**1:00**	**4:46**	**AM** ________
					PM ________
					TOTAL ________

12. 8:01 10:51 1:30 4:19 AM ________
PM ________
TOTAL ________

13. 6:58 11:02 12:00 4:00 AM ________
PM ________
TOTAL ________

14. 7:00 11:30 12:15 4:30 AM ________
PM ________
TOTAL ________

15. 8:06 11:53 12:31 3:08 AM ________
PM ________
TOTAL ________

16. Hallie clocked in at her job at GreatHair Totally at 8:10 AM. She worked 4 hours 20 minutes and then clocked out. What time did Hallie clock out? ______________

17. Bartholomew works as a hair stylist. He started his day at 9:03 AM and didn't leave until 7:55 PM If he took a one hour lunch break, what is the elapse time of actual work Bartholomew did on this day? ________________

18. During the day, Greta worked the following hours at her job at the spa business.
7:45 AM to 11:35 AM; 12:35 PM to 3:33 PM, and 4:15 PM to 6:10 PM

What was the total number of hours and minutes Greta worked? _____________

19. Allie did the late shift at the Health and Beauty Warehouse. She started at 12:20 AM and finished at 10:40 AM. What was the elapse time of her work day? ___________

20. Doug ended his day at 5:15 PM. If he worked 3 hours and 45 minutes straight, when did he start his workday? _______________

21. SallyJo started working at 8:30 AM and worked 3 hours 10 minutes, then after lunch, she started work at 1:15 PM and worked 3 hours 55 minute.
What was the clock time when she stopped before lunch? _________

What time did she stop at the end of the day? ________________

What was the elapse total from her start time to end time? ______________

4–3 ADDING HOURS AND MINUTES

Often in one's business in order to figure if appointment time durations can fit into a day schedule, a professional needs to be able to add different elapse times of hours and minutes. To do so

- First add the total number of minutes;
- If the total minutes is less than 60, then go on and add the total number of hours;
- If the total minutes are 60 or more, subtract 60 from the total for the number of minutes, and if the resulting minutes are less than 60, add 1 to the number of hours, then add the total number of hours; and
- If the resulting minutes is still more than 60, keep subtracting 60 from the minutes and add 1 to the hours each time, until there is less than 60 minutes, then add the number of hours.

For example,

Add elapse times 5 hours 12 minutes and 3hours 20 minutes

5: 12	
+3: 20	12 + 20 = 32 minutes. As this is less than 60, so then add 5 + 3 = 8 hours
8: 32	so combined 8 hours 32 minutes

Add elapse times 2 hours 45 minute and 3 hours 23 minutes

2: 45	
+3: 23	
5: 68	45 + 23 = 68 minutes, as this is more than 60, subtract 68 – 60 = 8 minute
	So, add 1 to hours
5: 68	becomes (5+1) : 08 = 6:08 so combined elapse time of 6 hours 8 minute

Add elapse times 3 hours 50 minute, 2 hours 50 minutes, and 30 minutes

3: 50	
+2: 50	
+: 30	
5: 130	add 50 + 50 + 30 = 130 minutes, which is more than 60, subtract 130 – 60 = 70 minutes, so add 1 to hours, However, 70 is more than 60, so subtract 70 – 60 = 10
	Minutes, so add 1 more to hours.
5: 130	becomes (5 + 2): 10 = 7: 10 so combines elapse time is 7 hours 10 minutes

Example:

Dollie has a beauty shop. She wants to figure out if the following appointments can fit into a four-hour afternoon time window, with a five-minute break between each appointment.

Hair treatment: 1 hour 15 minutes;

Facial: 55 minutes;

Pedicure: 32 minutes;

Hair styling: 1 hour 17 minutes;

For the four appointments, needed are three breaks of five minutes each.

We need to add these hours plus three breaks of 5 minutes (or a total of 15 minutes).

```
  1: 15
+ : 55
+ : 32
+1: 17
+ : 15
------
  2: 134
```

add 15 + 55 + 32 + 17 +15 = 134, which is more than 60, so 134 – 60 = 74

so add 1 Hour to the hours, and 74 is greater than 60, so 74 – 60 = 14,

so add 1 more hours to the hours

2:134 = (2 + 2):14 = 4:14 so combined 4 hours 14 minutes elapse time.

Therefore, Dollie would not be able to fit these four appointments with three breaks in the four-hour window.

Practice

Find total elapse time for

a) 3 hours 24 minutes and 6 hours 15 minute ________________

b) 10 hours 55 minute and 4 hours 7 minutes ________________

c) 3 hours 2 minutes and 12 hours 4 minutes ________________

d) 9 hours 55 minutes and 4 hours 15 minutes ________________

e) 3 hours 40 minute and 5 hours 10 minutes and 45 minutes ________________

f) 7 hours 11 minutes and 9 hours 49 minutes and 6 hours ________________

g) Ferdie needs to know the elapse time for the following jobs in her shop

Massage 2 hours 15 minutes, Haircut 50 minutes, Nail treatment 1 hour 20 minutes ____________________

h) Can Horatio fit the following treatments in an eight-hour time schedule? And if so, how much extra time does Horatio have for another appointment?

Three Hair stylings at 1 hours 20 minutes each

One Massage of 1 hour 15 minutes

One Shampooing of43 minutes

One Pedicure of 1 Hour 12 minutes

Answers:

a) 9 hours, 39 minutes;

b) 15 hours 2 minutes ;

c) 15 hours 6 minutes ;

d) 14 hours 10 minutes ;

e) 9 hours 35 minutes ;

f) 23 hours;

g) 4 hours 25 minutes;

h) yes, 7 hours 10 minutes total, with 50 minutes left for appointments.

4-3 Assignments

SHOW ALL WORK!!!

Find the total time in hours and minutes

1. **6 hours 14 minutes and 8 hours 59 minutes** ________________
2. **10 hours 2 minutes and 2 hours 5 minutes** _______________
3. **13 hours 2 minutes and 12 hours 22 minutes** ______________
4. **6 hours 15 minutes and 14 hours 54 minutes** ________________
5. **112 hours 44 minutes and 34 hours 3 minutes** ______________
6. **4 hours 11 minutes and 4 hours 11 minutes** _________________
7. **3 hours 40 minutes, 5 hours 10 minutes, and 45 minutes** ______________
8. **7 hours 11 minutes, 9 hours 49 minutes, and 6 hours** _______________
9. **2 hours 10 minutes, 4 hours 18 minutes, and 6 hours 1minute** ______________

10. **1 hours 11 minutes, 1 hours 11 minutes, and 4 hours 48 minutes ____________**

11. **3 hours 5 minutes, 5 hours 22 minutes, 7 hours 56 minutes, and 2 hours 59 minutes ____________________**

12. **6 hours 30 minutes, 5 hours 15 minutes, 34 minutes, 6 hours 8 minutes, and 2 hours 59 minutes _______________**

13. **GertieSue needs to know the total time for the following jobs in her shop**

Massage: 1 hours 15 minutes, Haircut: 45 minutes, Nail treatment: 1 hour 10 minutes _______________

14. **How many total hours and minutes would Hank put on his time sheet if his jobs totaled**

Haircut: 55 minutes

Haircut: 1 hour 10 minutes

Hair styling: 2 hour 45 minutes

Facial: 45 minutes

____

15. **Can Jethro fit the following treatments in an 8-hour time schedule? And if so, how much extra time does Jethro have for another appointment?**

Two Hairstylings at 1 hours 5 minutes each

Two Massages at 1 hour 30 minutes each

Two Pedicures at 45 minutes each

___________ ______________________________

16. **Hallie needs to schedule the following four appointments in a 5-hour morning session. Figure out how much time does she have for breaks between the appointments?**

Facial: 1 hour 10 minutes,

Haircut: 45 minutes,

Nails: 55 minutes,

Facial: 1 hour 15 minutes

____

4–4 USING MINUTES AS FRACTIONS OF HOURS

CONVERTING MINUTES INTO HOURS WITH FRACTIONS

As scheduling is often involves time as fractions of hours, it is important we can convert a fraction of an hour into total minutes.

As there are 60 minutes in an hour, then a total of minutes divided by 60 will give those minutes as a fraction of an hour.

For a 30 minute appointment as a fraction of an hour, 30 minutes is equal to $\frac{30}{60} = \frac{1}{2}$ of an hour. So a 30 minute appointment is a $\frac{1}{2}$ hour appointment.

A 45 minute time elapse is $\frac{45}{60} = \frac{3}{4}$ of an hour.

Example:

For a 2 hour 20 minute appointment, turning the 20 minutes into a fraction of an hour we get $\frac{20}{60} = \frac{1}{3}$ hour. So 2 hours 20 minutes is 2 hours $+ \frac{1}{3}$ hours $= 2\frac{1}{3}$ hours.

Example:

Turing 89 minutes into hours, we get $\frac{89}{60}$ hours $= 1\frac{29}{60}$ hours.

Here is a breakdown of fraction of an hour for 5 minute intervals

minutes	5 minutes	10 minutes	15 minutes	20 minutes	25 minutes	30 minutes	35 minutes	40 minutes	45 minutes	50 minutes	55 minutes
fraction of hour	1/12	1/6	1/4	1/3	5/12	1/2	7/12	2/3	3/4	5/6	11/12

CONVERTING HOURS WITH FRACTIONS INTO MINUTES

When we have a fraction of an hour, we need to turn the fraction into an equivalent fraction with a denominator of 60, and the numerator will be the number of minutes.

For example,

An appointment is scheduled to last $3\frac{2}{3}$ hours, and we want the time in hours and minutes. We need to find the number of minutes in 2/3 of an hour. To do so, we create a proportion equation, with × = number of minutes in the hour.

$$\frac{2}{3} = \frac{x}{60}$$

To solve for x, we cross multiply to get $2(60) = 3x$ or

$120 = 3x$ divide both sides by 3

$$\div 3 \quad \div 3$$

$$40 = x$$

So 2/3 of an hour is 40 minutes. And $3\frac{2}{3}\,hours$ = 3 hours 40 minutes.

Let's say we have an elapse time of $6\frac{21}{30}$ hours, and we want to figure this time in a total of minutes. We can do so in two ways:

- Way 1 , We can turn $\frac{21}{30}$ hour into minutes by figuring

$$\frac{21}{30} = \frac{x}{60}$$

To solve for x, we cross multiply $60(21) = 30x$ or

$1260 = 30x$ divide both sides by 30

$$\div 30 \quad \div 30$$

$$42 = x$$

So, 21/30 of an hour is 42 minutes. And, 6 hours is 6 (60 minutes) = 360 minutes.

So $6\frac{21}{30}$ hours = 360 minutes + 42 minutes = 402 minutes.

- Way 2, We turn $6\frac{21}{30}$ into an improper fraction

$$6\frac{21}{30} = \frac{(6 \bullet 30 + 21)}{30} = \frac{180 + 21}{30} = \frac{201}{30}$$

To turn this into minutes we figure $\frac{201}{30} = \frac{x}{60}$

To solve for x, we cross multiply $201(60) = 30\,x$ or

$12060 = 30\,x$ divide both sides by 30

$$\div 30 \quad \div 30$$

$$402 = x$$

So $6\frac{21}{30}$ hours = 402 minutes.

Practice:

a) Turn elapse time into hours with fraction

50 minutes________ 1 hour 14 minutes ______ 4 hours 21 minutes ______ 326 minutes ______

b) Turn hours in total minutes

$\frac{1}{4}$ hours ______ $3\frac{11}{30}$ hours ______ $17\frac{11}{12}$ hours ______ $4\frac{59}{60}$ hours ______

Answers:

a) $\frac{5}{6}$ hour; $1\frac{7}{30}$ hour; $4\frac{7}{20}$ hours ; $5\frac{13}{30}$ hours

b) 15 minutes; 202 minutes; 1075 minutes; 299 minutes

More practice:

Add the following hours and give answer in hours as fraction and in hours and minutes

$3\frac{2}{3}$ *hours plus* $4\frac{2}{5}$ hours ______________ ______________

Answer: $3\frac{2}{3} + 4\frac{2}{5} = 3\frac{10}{15} + 4\frac{6}{15} = 7\frac{16}{15} = 8\frac{1}{15}$ hours ; $8\frac{1}{15}$ hours = 8 hours 4 minutes

Add the following hours and give answer in hours as fraction and in hours and minutes

$2\frac{3}{4}$ *hours*, $5\frac{1}{8}$ *hours*, $6\frac{3}{20}$ hours , $2\frac{1}{3}$ *hours* ____________ ____________

Answer: $2\frac{3}{4} + 5\frac{1}{8} + 6\frac{3}{20} + 2\frac{1}{3} = 2\frac{90}{120} + 5\frac{15}{120} + 6\frac{18}{120} + 2\frac{40}{120} = 15\frac{163}{120} = 16\frac{43}{120}$ hours

$16\frac{43}{120}$ hours = 16 hours $21\frac{1}{2}$ minutes

KellySue recorded the following hours of work for different jobs in the Beauty shop during the week of June 6 – June 11

date	6-Jun	7-Jun	8-Jun	9-Jun	10-Jun
	Haircutting 1 ¼ hour	Haircutting 1 ¼ hour	Facial 1 1/8 hour	Haircutting 1 ¼ hour	Facial 2 1/8 hour
	Haircutting 1 ¼ hour	Hair styling 1 ¾ hour	Haircutting 1 ¼ hour	Haircutting 1 ¼ hour	Haircutting 1 ¼ hour
	Haircutting 1 ¼ hour		Haircutting 1 ¼ hour	Haircutting 1 ¼ hour	Haircutting 1 ¼ hour
	nail appt 2/5 hour		Nail appt 2/5 hour	nail appt 4/5 hour	Nail appt 3/5 hour

Give the total hours in hour as fractions and in hours and minutes by date

June 6 ____________ ____________ June 7 ________ ________

June 8 _________ ____________ June 9 ________ _________ June 10 _________ _________

And, give the total hours in hour fractions and in hours and minutes for total week.

_____________ __________________

Answer: June 6: $4\frac{3}{20}$ hr ; 4 hours 9 minutes ; June 7: 3 hours; 3 hours ; June 8: $4\frac{1}{40}$ *hr* ; 4 hours $1\frac{1}{2}$ mintue

June 9: $4\frac{11}{20}$ hr; 4 hours 33 minutes; June 10: $5\frac{9}{40}$ hr ; 5 hours $13\frac{1}{2}$ *minutes*

Total for the week $20\frac{19}{20}$ hours ; 20 hours 57 minutes

4-4 Assignments

Turn elapse time into hours with reduced fraction. Show all work.

For example, 2 hours 34 minutes = $2\frac{34}{60}$ *hours* $= 2\frac{17}{30}$ *hours*

1. 6 hours 33 minutes ______________
2. 1 hour 1 minute _______________
3. 6 hours 59 minutes _______________
4. 2 hours 10 minutes _______________
5. 19 hours 24 minutes _______________
6. 345 minutes ______________
7. 121 minutes ______________
8. 676 minutes ________________
9. 3600 minutes ______________
10. 1 hour 3 minutes __________

Turn hours into hours and minutes

For example $3\frac{1}{6}$ *hour* = 3 hours 10 minutes

11. $7\frac{5}{12}$ *hours* ___________

12. $1\frac{1}{60}$ *hours* __________

13. $\frac{7}{12}$ *hour* __________

14. $9\frac{7}{8}$ *hours* __________

15. $112\frac{3}{16}$ *hours* __________

Turn hours into total number of minutes

For example $5\frac{3}{5}$ *hours* = 336 minutes

16. $2\frac{2}{3}$ *hours* __________

17. $12\frac{3}{8}$ *hours* __________

18. $4\frac{11}{12}$ hours __________

19. $220\frac{13}{15}$ hours __________

20. $7\frac{37}{60}$ *hours* __________

Add the following hours and give answer in hours as fraction and in hours and minutes

For example $1\frac{2}{3}$ *hours plus* $2\frac{3}{5}$ hours is $4\frac{4}{15}$ hours and 4 hours 16 minutes

21. $3\frac{3}{4}$ *hours and* $5\frac{3}{4}$ *hours* __________ __________

22. $2\frac{1}{2}$ *hours and* $11\frac{13}{15}$ hours __________ __________

23. $7\frac{1}{3}$ *hours and* $9\frac{5}{6}$ *hours* __________ __________

24. $4\frac{1}{4}$ *hours and* $5\frac{7}{12}$ hours and $2\frac{7}{10}$ hours __________ __________

25. $11\frac{1}{6}$ *hours and* $\frac{3}{4}$ *hour and* $10\frac{4}{15}$ *hours* __________ __________

26. $2\frac{1}{2}$ *hours and* $4\frac{1}{8}$ *hours and* $6\frac{2}{3}$ *hours and* $\frac{3}{10}$ *hour* ______ _____

27. Calvin logged the following time for jobs done on Monday.

$Manicure: 2\frac{1}{2} hours;\ Facial: 1\frac{1}{5} hours;$

$Hair\ Treatment: 1\frac{7}{8} hour; Pedicure: \frac{5}{6} hour;$

Haircutting: $1\frac{1}{4}$ *hours* ; Haircutting: $1\frac{1}{8}$ hour

Calvin is paid extra hourly pay for any time logged in excess of $7\frac{1}{2}$ hours in a day.

a) Figure the total hours logged in hours as fraction and in hours and minutes

____________________ ____________________

And b) how much time in hours as fraction and in hours and minutes will Calvin get extra pay

____________________ ____________________

28. LouisaMae recorded the following hours of work for different jobs in the JustGorgeousBeauty shop during the week of July 6–July 11

date	6-Jul	7-Jul	8-Jul	9-Jul	10-Jul
	Haircut 1 hour	Haircut 1 ¼ hour	Facial 1 1/8 hour	Haircut 1 ¼ hour	Facial 2 1/8 hour
	Haircut ¾ hour	Hair dying 2 ¾ hour	Haircut 1 hour	Haircut ¾ hour	Haircut 1 ¼ hour
	Facial 1 1/8 hour	Pedicure 1 ¼ hour	Haircut 1 ¼ hour	Haircutting 1 ¼ hour	Haircut 1 ¼ hour
	Pedicure 1 2/5 hours		Nail appt 3/5 hour	nail appt 4/5 hour	Nail apt 3/5 hour
	Hairstyling 2 ¼ hours			Hairstyling 2 ¼ hour	Hair treatment 2 hours

Give the total hours in hour as fractions and in hours and minutes by date

July 6 __________ ________ July 7 _______ _______ July 8 ________ _________

July 9 __________ ________ July 10 _________ _________

In addition, give the total hours for the week in hour fractions and in hours and minutes for total week.

_______________ __________________

CHAPTER 5

Pricing Products and Services, Tax on Products, and Determining Profit

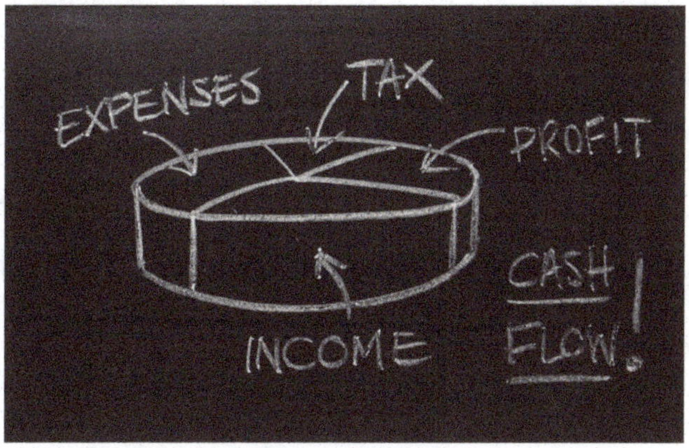

If you plan is to run a cosmetology/esthetic business, it is important to understand that to stay in business and to be successful you must be familiar with your costs or expenses, sales or revenues, and profit. Moreover, to remain successful long term, you must eventually earn a continuous **positive profit**. So understanding what profit is and how it is calculated is very important. This involves solving for unknown quantities through formulas. The more you know about how to handle the math using formulas the better to become in handling the kind of problems involving sales, costs, discounts, markups, tax, and profits.

5-1 CALCULATING SALES TAX

When items are purchase or are sold in your business, taxes on the price that goes to the government are usually part of the calculation. We call this SALES TAX.

Given T = tax, P = price, R = tax interest rate, The Formula to figure tax is tax = price times tax interest rate or

$$\mathbf{T = P \times R}$$

As tax is a dollar amount, we figure tax to the nearest hundredth (or nearest cent).

Also, if taxes are involved in a purchase or a sale, given F = the final cost or price of items purchase or sold, P = original price, and T = Tax, then the formula for final cost or price of items purchase or sold = original price + Tax or

$$F = P + T$$

For example, Gertie's Gorgeous Look Store buys $250 of Hair Cleaning Supplies. Local regulations have a 5% tax on any purchase. What is the tax amount of Gertie's purchase, and what is the final cost?

The formula to figure tax is: $T = P \times R$ with P = the price in dollars, R is the tax rate, and T = tax in dollars.

For our problem P = $250, r = 5% = .05. Then for $T = P \times R$ using substitution of P and R,

$T = 250 \times .05 =$ $12.50. The sales tax is $12.50.

With F = final cost or price final cost the formula is F = P + T, so F = 250 + 12.50 = $262.50. The final cost to Gertie's is $262.50

Example

FrannieMae sold bottles of shampoo at $21.00 a bottle. The sales tax rate on the price is 8.8%. What is the sales tax on a bottle of shampoo? And what is the final cost to the customer?

P = $21.00, R = 5.5% = .055, and tax is T = P x R = $21.00 x .055 = $1.16

And the final cost is F = P + T = $21.00 + $1.16 = $21.16

SHORTCUT: Another way to figure the final cost or price given the tax rate is to add the tax rate to 100% then multiply by the cost or price. Using this in our calculations, we change the tax % rate to a decimal, add the decimal to 1, the multiply by the cost or price.

For example, what is the final price of a $40.00 purchase at a 6% tax rate?

To figure the final price, we add the purchase price and the 6% sales tax on the purchase price

or $40.00 + .06 × $40.00 = $40.00 + $2.40 = $42.40

This can also be figured as

(1) ($40.00) + (.06) ($40.00) = (1 + .06) ($40.00) = (1.06) ($40.00) = $42.40

You should see we add .06 to 1 then multiply by $40.00 to get the final cost of $42.40. Then to figure the sales tax by

$42.40 – $40.00 = $2.40 in sales tax

Example:

Find the final cost of a $250 cost at 7% sales tax,

(1 + .07) ($250) = (1.07) ($250) = $267.50

And that means \$267.50 – \$250 = \$17.50 in sales tax

Example:

Find the final price of a \$62.50 bottle of perfume at 9.9% sales tax.

$$(1 + .099)(\$62.50) = (1.099)(\$62.50) = \$68.69$$

And that means \$68.69 – \$62.50 = \$6.19 in sales tax.

What if we know the price of an item and the sales tax amount and want to figure the sales tax rate? By manipulating the sales tax formula, we can "solve" for R in terms of P and T. As T = P × R, we divide both sides by P to get

$$T/P = R$$

For example, the sales tax is \$2.36 on an item for sale that is priced at \$29.50. What is the sales tax rate?

T/P = R, so \$2.36 / \$29.50 = .08, so the tax rate is 8%

What if we know the sales tax amount and the sales tax rate on an item and want to figure the price? By manipulating the sales tax formula, we can "solve" for P in terms of T and R.

As T = P × R, we divide both sides by R to get

$$T/R = P$$

For example, on an item for sale, the sales tax rate is 3%, and the sales tax is \$28.80. What is the price of the item?

T/R = P, so \$28.80/.03 = \$960.00, so the price is \$960.00

Practice

a) Find the tax and final price on a \$50 haircut that carries a 6.2% sales tax. ______ ______

b) Lonnie buys a \$2350 Hair Styling Chair with a 7.25% sales tax. Find the sales tax and final price on the item. ______ ______

c) Reginald sells sets of decorative fake nails at \$7.99 each. The sales tax is 4.66%. What is the sales tax and the final price on this item? _______ _______

d) Fabio's Famous Parlor buys Facial cream at \$230 a box. There is a sales tax of 8.88%. What are the sales tax and final cost of the carton? _____ ______

e) The sales tax on a bottle of shampoo was \$3.48 for a list price of \$14.50 a bottle. What is the sales tax rate? ___________

f) The sales tax rate is 4.6% and the sales tax is \$3.91 on a makeup kit. What is the selling price of the makeup kit? _______

Answers

a) \$50.00 x .062 = \$3.1; \$50.00 1.062 = \$53.10

b) \$2350.00 x .0725 = \$170.38; \$2350.00 x 1.0725 = \$2520,38

c) \$7.99 x .0466 = \$0.37; \$7.99 x 1.0466 = \$8.36

d) \$230.00 x .0888 = \$20.42; \$230.00 x 1.0888 = \$250.42

e) \$3.48/\$14.50 = .24, so 24% sales tax rate

f) \$3.91/ .046 = \$85.00

5-1 Assignments

Show all work steps!

1. Paula sold sets of the fake nail at \$33 a set at her salon. Her city charges 6.6% sales tax. What is the price that Paula's customers pay for a set? __________
   What is the sales tax paid? ___________

2. De La Wonder Hair Salon sells hair coloring product at \$22.00 a box. A 10.3% sales tax is applied. What is the actual cost for a box to the customer? ___________
   What is the sales tax paid? _________

3. Barney's Beauty sells makeup mirrors at \$21.50 each. The sales tax rate is 9.99%. What is the total price customers pay for a mirror? _________
   What is the sales tax paid? _______

4. Felicia Hair Salon bought a Styling Chair at a list price of \$880.00 from a wholesaler. There was a 7.5% sales tax. What does Felicia pay to the wholesaler for the chair? ________
   How much of that payment is sales tax? _______

5. Nail polish selling price is \$9.90 a bottle. If there is a 5% sales tax, what does the customer pay for a bottle of nail polish? __________
   How much of that payment is sales tax? _____

6. On a \$45.00 purchase from beauty store, there was a \$3.60 sales tax applied. What is the sales tax percentage rate? ________

7. The 9.00% sales tax on a bottle of conditioner came out to \$1.08. What is the selling price of the conditioner? _________
   What does a customer pay for the bottle? ________

8. A customer at Sara's Spectacular Spa bought the following items at selling price as follows
Bottle of moisturizer, $15.50

   Bottle of Home mist $9.90

   Bath oil salts $12.50

   Jar of Facial Scrub $7.60

   If the sales tax for these items came out to $4.55, what is the sales tax rate? _______

9. Randy sells liquid hand soap at his boutique at a selling price of $5.60 a bottle. If there is 5.5% sales tax, what does a customer pay in taxes for a bottle? ________

10. On a manicure table from a distributor, sales tax of 11.20% came out to a sales tax of $39.20. What is the list selling price for the table? ________

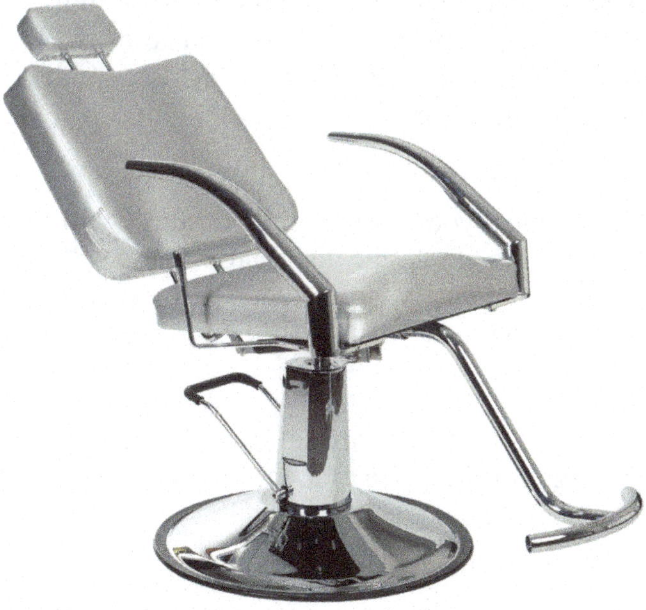

11. Dolores shopped at HoneyBee's Beauty and Boutique and bought three bottles of Shampoo at a selling price of $12.50 per bottle and two bottles of body wash at a selling price of $9.50 per bottle. If sales tax for the items came out to $6.50, what is the % sales tax rate to the nearest tenth of a percent? __________

12. LouiseMae's Hair Salon charges $16.50 a haircut. If sales tax is 10.5%, what does a customer pay for a haircut? ___________

13. Fantastic Facial Products had face cream selling at $34.00 a tube. There is a 4% local tax and a 5% sales tax applies to the tube. What is the price a customer pays for the tubes? __________

14. Harry's Hair bought a barber chair from WholeWorld Beauty at a list price of $999. There is an 8.7% sales tax. What is Harry's cost for the chair? ________

15. On a $34.55 selling price for a nail set, the sales tax came out to $3.80. What was the sales tax percentage? _________

5-2 MARKUPS

In your beauty business, you may buy goods or products from distributors, wholesalers, or other retailers that you will then sell to your customers. In order to make a profit on those goods and products that you sell to your customers, a markup needs to be added to the cost you incur.

A MARKUP is a difference between the cost of a good or service and its selling price.

If cost = C, S = Selling price, and M = markup, then

$$C + M = S$$

Markup can be expressed as a fixed amount or as a percentage of the total cost or selling price. When a markup is a percentage of the cost, with C = cost, R = rate of markup, and M = markup, then

$$M = R \times C$$

Many times, your cosmetology business will buy goods and products direct from a manufacturer or through a wholesale distributor. These selling businesses will often sell their product at a lower price as an incentive to get your business. You will want to mark up the products in order to earn a profit on the sale.

Example:

Bailey's Beauty Emporium used a markup rate of 40% on the cost of a bottle of conditioner. Find the selling price of a bottle of conditioner that cost the Bailey's $25 from the beauty goods wholesaler.

The markup is 40% of the $25 cost, so $\underline{R}$ = .40, C = $25, and the markup is:

M = R × C = (.40) ($25) = $10. So the markup is $10.

Then the selling price S = C + M = $25 + $10 = $35, so the conditioner sold for $35.

Example:

Seallie's Hair Salon buys nail polish from a wholesaler for $450.00 a carton. If there are 100 bottles in a carton, and Seallie marks up the bottles 35%, what will they sell the bottles for?

$450 divided by 100 = $4.50, so the cost for Seallie is $4.50 a bottle

The markup is 35%, so R = .35, C = $4.50 and M = R × C = (.35) ($4.50) = $1.58 And the Selling price S = M + C = $1.58 + $4.50 = $6.08, so the bottle sells for $6.08.

SHORTCUT: Another way to figure markups is to add the markup percentage rate to 100% then multiply by the original cost. Using this in our calculations, we change the markup % rate to a decimal, add the decimal to 1 , then multiply by the original cost.

For Example, a 22% markup is applied to a $195 cost. This is the cost plus 22% of the cost or

195.00 + (.22) ($195.00) = $195.00 + $42.90 = $ 237.90 selling price

We can think of this as:

(1)($195.00) + (.22) ($195.00) = (1 + .22) ($195.00) = (1.22) ($195.00) = $237.90

So we add .22 to 1 then multiply by $195.00 to get the selling price of $237.90 And since M + C = S, then M = S – C so the markup amount is $237.90 – $195.00 = $42.90

Example:

The Allure Store buys facial cream from a wholesaler at $11.75 a container. They markup the facial cream 65%. What is the selling price to their customers?

65% markup means (1 + .65) ($ 11.75) = (1.65) ($11.75) = $19.39 selling price a container, and the markup is $19.39 – $11.75 = $7.64

Example:

Jason buys fingernail cleaner from the distributor for at $6.75 a bottle. He marks up the cleaner 100% to sell in his shop. What is the final selling price?

100% markup means (1 + 1) ($6.75) = 2($6.75) = $13.50 a bottle, and the markup is $13.50 – $6.75 = $6.75.

What if we know the price and the markup amount and we want to figure the markup % rate?

We manipulate our markup formula ***M* = *R* × *C*** to solve for R in terms of M and C by dividing both sides by C to get

M/C = R

Example:

On a list price of $12.60 a bottle of shampoo, a beauty store marks up the selling price by $10 a bottle. What is the % markup?

As **M/C = R** then $10.00/$12.50 = .80, so the markup is 80%

What if we know the markup and the markup % rate and we want to figure the original price? We manipulate our markup formula **M = R × C** to solve for C in terms of M and R by dividing both sides by R to get

M/R = C

Example:

The percentage markup of 95% came out to a $23.00 markup on a list price for a product sold. What is the original list price and the final selling price?

As **M/R = C** then $23.00/.95 = $24.21 list price.

MULTIPLE MARKUPS

Sometimes products or services are marked up more than once. Each markup is figured on the current price at the time of the markup. For example, a product is purchased for $5.50 to sell in a store. It is marked up 65%, then a week later is marked up an additional 40%. We figure the initial markup as (1 + .65) ($5.50) = (1.65)($5.50) = $9.08.

We then use $9.08 as the "current" price to apply the additional 40% or

(1 + .40) ($9.08) = (1.40) ($9.08) = $12.71 So, the final selling price is $12.71

Note, as $9.08 = (1.65) ($5.50), then (1.40) ($9.08) = (1.40) (1.65) ($5.50) = $12.71

So we can figure the final selling price for multiple markups by multiplying the individual markups by the original price. And the order of the markup being multiplied doesn't matter.

Example:

A bottle of Shampoo was purchased by a Salon for $18.00, then marked up 25%, followed later by an additional markup of 30%.

25% markup is 1.25 of the price and 30% is 1.30 of the price.

The final selling price is (1.25) (1.30) ($18.00) = $29.25. This would be the same result is the 1.25 and 1.30 were rearranged, as (1.30) (1.25) ($18.00) = $29.25

MARKUPS AND SALES TAX

When applying sales tax to an item that has been marked up, you apply the sales tax on the **final** selling price, so after the markup.

Example:

A jar of Hair Gel bought from a wholesaler for $7.50 a jar is marked up 55% by Greta's Beauty store. There is a 6% sales tax. What is the cost to the customer for a jar of the gel?

First figuring the selling price, (1 + .55) ($7.50) = (1.55) ($7.50) = $11.63.

Second figure the sales tax on the $11.70, (1.06)($11.70) = $12.33. So the customer pays $12.33.

Just like multiple markups, a markup followed by a sales tax % can be done in one continuous calculation as

$7.50 (1.55) (1.06) ($7.50) = $12.33

Practice:

a) BoldBeauty Shoppe buys a special shampoo for $18.00 a bottle from the distributor. BoldBeauty Shoppe marks up the soap 100%. What is the selling price of the bottle?

b) Wonderful Spa Experience used a markup rate of 85% on the cost of soothing oils. Find the selling price of the oils that cost the retailer $20.50 a bottle from the beauty goods wholesaler.

c) HairAffair Salon buys Conditioner from a wholesaler for $ 550 a carton. If there are 50 bottles in a carton, and Hair Affair marks up the bottles 65%, what will they <u>sell each of the bottles</u> for?

d) Polly's Pamper Salon bought hand lotion at $7.50 a bottle from a distributor. At first, Polly marks up the lotion 95%. After one month, Polly mark up the lotion an additional 15%. What is the final price for a bottle of the lotion?

e) Helene Hair Store marks up a hair product 110% which came out to a $22.00 markup. What was the list price before the markup? What was the final selling price?

f) A list price of $30.00 on a beauty item for sale was markup $18.00 for a final selling price. What was the % markup?

g) Yvette bought a $6.60 bottle of facial soap from her distributor, then marked it up 85% to sell at here store. If there is a 7.5% sales tax, what does a customer payout for a bottle of the facial soap at Yvette's store?

Answers

a) 2.00 x $18.00 = $36.00 a bottle

b) 1.85 x $20.50 = 37.93 a bottle

c) $550.00/50 = $11.00 list cost per bottle; 1.65 x $11.00 = $18.15 a bottle

d) 1.95 x 1.15 x $7.50 = $16.82 a bottle

e) $22.00/1.10 = $20.00 list price. $22.00 + $20.00 = $42.00 selling price

f) $18.00/$30.00 = .60 so a 60% markup.

g) (1.85) (1.075) ($6.60) = $13.13 a bottle

5-2 Assignments

Show all work!!

1. Delightful Body Works buys hand cream at $15.50 a container. The want to sell the containers to their customers with a markup of 55%. What is dollar amount of the markup for a container? __________
What would the selling price for the container be at the store? ___________

2. Carpe Diem Cosmetics sells hand lotion. The bottles of lotion they bought from a wholesaler at $11.30 a bottle are marked up 100%. What is the selling price for a bottle in the store? ______

3. Lola Hair bought hair cream at $14.00 a tube from Beauty Suppliers. Lola marks up the hair cream 75%. What is the dollar amount of the markup? _________

4. Bath and Beauty Shoppe buys a special bubble bath soap for $11.00 a bottle from the distributor. Bath and Beauty Shoppe marks up the soap 85%. What is the selling price of the bottle? ____________

5. Beauty "R" US store used a markup rate of 55% on the cost of Stainless Steel Slant Tweezers for sale in their store. Beauty "R" US paid $38.00 for the Tweezers from the beauty goods wholesaler. Find the selling price of the tweezers ____________
How much in dollars was the markup per tweezer? _______

6. Wonderful Hair Salon buys hair crème from a wholesaler for $250.00 a carton. If there are 100 tubes in a carton, and Wonderful marks up the bottles 45%, what will the <u>selling price per tube</u> be? ______________

7. Gorgeous Guys and Gals bought massage lotion at $12.50 a bottle from a distributor. At first, Gorgeous marks up the lotion 35%. After one month, the store marks up the lotion an additional 10%. What is the final selling price for bottle of the lotion? ______________

8. ShellyMae's, a beauty store, bought UltraRich Conditioner at $16.50 a bottle from the distributor. ShellyMae marked up the bottles for sale 100%. If there is a 7% sale tax, what is the price for the customer for a bottle of the conditioner? _______

9. Frannie's Nails bought sets of nails for $25 a box, then marked up the boxes 35% for sell. If there is an additional sales 10% taxes on the boxes, what does the customer pay to purchase a box of nails? _________

10. Rollie sells scented candles at her boutique. She bought them at $9.95 per candle from her supplier then marked them up 45%. A week later she marked then up an additional 5%. What was the final selling price to the customer?________

11. The Soothing Spa stocked body oils for sale. The bought oils for $15.78 a bottle from a distributor and then marked the bottle up by 95%. The after a few weeks, the spa marked up the oils another 22%. What was the final selling price to the customer? _________

12. Retta's Radical Beauty bought makeup kits at $7.88 each from there wholesaler, then added a $9.46 markup to the price. What is the markup % for the kits? ______

13. Petunia for her salon purchased shampoo 24 bottles in a case from Terrific Hair Product Distributors at $360 a case. At first, Petunia marked up the bottles 75%, then later an additional 25%. If there is a 6% sales tax , what is the final cost to her customer for a bottle?

____

14. A 65% markup on a item for sale at health store was a total markup of $10.92. What was the original list price before markup?

____

What was the final selling price? __________

15. A final selling price for a bottle of skin smoother was $28.47 a bottle. This included a markup of the original list price of 95%. What was the original list price? __________
What was the dollar amount of the markup? __________

5-3 DISCOUNTS OR MARKDOWNS

When you sell items or services to your customers you often will offer discounts to make the products more desirable to buy and to decrease inventory. In the same way, when you buy products and supplies, discounts can be offered to you as incentives to use their products. Discounts are also referred to as markdowns.

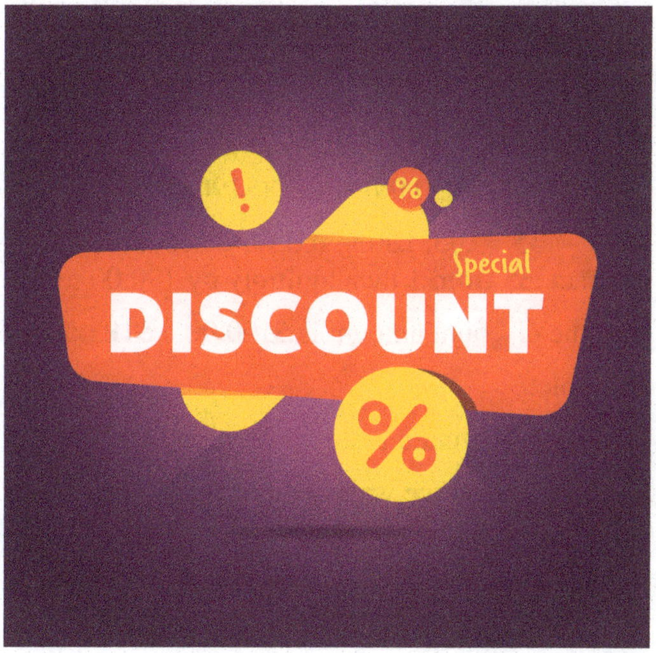

A discount is a dollar amount that is taken off the price or cost.

So if New Price = N , D = discount and P = original price,

$$N = P - D$$

When discounts appear as a percentage, we need to figure the percent of the original price or cost and subtract it from the price or cost. The formula for discount or markdown, given D for discount in dollars, R for discount rate, P for Price is

$$D = R \times P$$

Let's look at some examples:

Example:

CutiePie Cosmetics want to apply 15% markdown off the selling price of their skin lotion for a Holiday sale. The lotion has been selling for $17.00 a bottle. What is the new selling price after the markdown?

The formula for discount or markdown, given D for discount in dollars, R for discount rate, P for Price is D = R x P

P = $17.00. As 15% needs to be taken off the price, R = .15.
So the Discount is 15% of $17.00 or D = R x P = .15 x $17.00 = 2.55. So the discount is $2.55 off the original Sales price.

The formula for the New Sales price , letting N = new price, is N = P – D
Or N = $17.00 – $2.55 = $14.45
So the new bottle sells for $14.45.

Example:

Your beauty business buys nail polish from a distributor. You want to order $105.00 of polish, and the distributor, as an incentive, offers you a 10% discount.

So P = $105.00, R = .10 , and D = R x P = .10 x $105.00 = $10.50, so the discount is $10.50.
The New cost to you is N = P – D = $105.00 – $10.50 = $94.50 or you pay $94.50

SHORTCUT: Another way to figure a new discounted price or cost is to subtract the percent discount from 100% then multiply by the price or cost. To calculate this, we change the % rate to a decimal, subtract it from 1, then multiply by the price or cost.

For example, find the new discount price for a $190.00 item marked down 35%.

This is the original price minus 35% of the original price or

$190.00 – (.35)($ 190.00) = $190.00 – $66.50 = $ 123.50 new price

We can think of this as:

(1)($190.00) – (.35)($190.00) = (1 – .35) ($190.00) = (.65)($190.00) = $123.50

As 1 – .35 = .65, taking a 35% discount is saying that we want 65% of the original price. So we subtract .35 from 1 then multiply by 190.00 to get the new price $123.50

And the discount amount is $190.00 – $123.50 = $66.50

Example:

Find the discounted price for an item listed at $78.88 then discounted 21%.

1 –.21 = .79, so we want 79% of the list price, and

(1 – .21) ($ 78.88) = (.79) ($78.88) = $62.32 new price
And the discount amount is $78.88 – $62.32 = $16.56

Example:

A Beauty wholesaler marks down an item for sale listed at $410.20 by 32%. What is the final price to the buyer?

1 – .32 = .68 so we want 68% of the list price, and

(1 –.32) ($410.20) = (.68)($410.20) = $278.94 final price

And the mark down is $410.20 – $278.97 = $131.26

Example:

Mineral Springs Spa wants to offer a Spring Spa Special at 20% off to attract more customers.

A Spa Special costs $115.50.

1 – .20 = .80 so we want 80% of the list price , and

(1 – .20) ($115.50) = (.80)($115.50) = $92.40 final price

And the discount is $115.50 – $92.40 = $23.10

What if we know the list price and the amount of the markdown and we want to figure the % markdown. We manipulate the markdown formula D = R x P to solve for R in terms of P and D by dividing both sides by P. We get

D/P = R

Example:

A list price of $50.00 on an item is marked down $18.00. What is the markdown % rate for this mark down?

As D/P = R, then $18/$50 = .36 so a 36% markdown

What if we know the markdown amount and the markdown % rate and we want to figure the original list price. We manipulate the markdown formula D = R x P to solve for P in terms of D and R by dividing both sides by R. We get

D/R = P

Example:

A 25% discount on a product for sale amounts to $30 in markdown dollars. What is the original selling price of the product?

As D/R = P, then $30 / .25 = $120 original List price.

MULTIPLE DISCOUNTS OR MARKDOWNS

Sometimes, there is a need to discount services or goods more than once.

You figure the first discounted cost or price, then you use that cost or price to figure the second discounted cost or price, then use that one to figure the next, and so on.

For example,

Good Beauty Shoppe sells a special shampoo for $35.00 a bottle. After a month, the Shoppe markdown the remaining bottles 10% to try to get them sold. After another month, the few remaining bottles are discounted 25% more to clean out the inventory. What is the final price of these remaining bottles?

The $35.00 bottles are first markdown 10%, so the markdown price is

$$(1 - .10)\ (\$35.00) = (.90)\ (\$35.00) = \$31.50$$

Now, this new $31.50 price is marked down an additional 25%. The second markdown price

Is $$(1 -.25)(\$31.50) = (.75)\ \$(31.50) = = \$26.63.$$

So the final price of the remaining bottle is $26.63

Another way we can find this is by doing a continuous calculation, like this…

As the $31.50 = (.90)($35.00), (.75) ($31.50) = (.75)(.90)($35.00) = $26.63

Note:

The order of the two discounts doesn't matter, as (.75) (.90) ($ 35.00) = (.90) (.75) ($35.00).

So for multiple discounts, we find this by multiplying the two (or more) markdowns times the original list price.

Example:

Paulie's Perfect Hair has a shampoo and sets special for $40.00. Paulie marks down the special after two weeks by 8%, then an additional 12% after a month. What is the final price?

As 8% markdown is 1 – .08 = .92 and 12% is 1 – .12 = .88, Then our final price is (.92) (.88) ($40.00) = $32.38 price.

MARKUPS AND MARKDOWNS

Sometimes, goods or products are marked up **then** marked down later in order to stimulate sales or reduce inventory. What about a markup followed by a markdown?

We figure the selling price after one markup. Then we use that selling price with the markdown to find the new selling price, and so on.

Example:

Frances Fantasticks Salon buys 100 jars of styling wax from the manufacturer for $12.00 a jar. She marks up the jars 25%. After a month she marks down the remaining jars 15% to stimulate sales.

What is the final selling price of the jars?

Applying the 25% markup we figure, (1 + .25) ($ 12.00) = (1.25)($12.00) =$15.00. So Frances originally sells the bottle at $15.00 each.

Now we apply the 15% markdown to the $15.00 price. We figure (1 – .15) ($15.00) = (.85) ($15.00) = $12.75.

So the final selling price is $12.75 a jar.

You can also continuously do this by figuring final selling price by multiplying markup times markdown times the original cost or

(1.25) (.85) ($ 12.00) = $ 12.75

Example:

Terry's Spa bought Spring line towels at $15.00 each from a distributor. Terry marks up the towels 60%. After one month, she marks down the remaining inventory 20%, then a final sale a month later at a mark down of 30%. What is the final selling price for a towel?

A 60% markup is (1 + .60) = 1.60; a 20% markdown is 1 – .20 = .80= .80; a 30% markdown 1 – .30 = .70. To find the final selling price we figure

(1.60)(.80)(.70)($15.00) = $13.44 the final selling price.

Example:

Joey's Barber Shop charged $15.00 per haircut. Joey marked up the price 20% after a few weeks, the marked the price down 30% a week later. If a customer pays for the final sales price for the haircut and there is a 5% sales tax, what does the customer pay?

A 20% markup is (1 + .20) = 1.20; a 30% markdown is (1 – .30) = .70; a 5% sales tax is (1 + .05) = 1.05.

So (1.20) (.70) (1.05) ($15.00) = $13.23 cost to customer.

Practice

a) Hand's Full of Beauty sells hand moisturizer at $13.50 a bottle. A month later, they marked down the hand cream 25% in order to get rid of the inventory. What is the new selling price?

b) Fancy Flo's Shoppe puts a sales price of fingernail polish at $8.50 a bottle for her customers. After several weeks, Flo marked down the tube price 22%. What is the new sales price of a tube?

c) Debbie sees ultra-precise tweezers listed at $22.70 per tweezer from her favorite beauty distributor. The distributor offers her a 25% trade discount. What did Debbie pay for the tweezers?

d) A $10.50 item for sale was marked down $2.50. What is the discounted % rate?

e) A $12.50 markdown represents a 15% markdown of an original price. What is the original price and what was the final selling price?

f) Simply Alluring Store sells makeup sets at $33.00 a set. After a month, the store marks down the set $20%, then after another month marked the set down another 15%. What is the final selling price of the markup sets?

g) Stunning Style Salon wanted to buy lipstick sets to sell in the store. The see a set they liked from the beauty wholesaler listed at $24.60 a set. The distributor offers a 10% trade discount. Stunning Style the marks up the set 65%, then after a few weeks, they markdown the set 18%. If there is a 5.5% sales tax on items sold, what would the final cost be to the customer per set?

Answers

a) .75 x $13.50 = $10.13 a bottle

b) .78 x $8.50 = $6.63 bottle

c) .75 x $22.70 = $17.03 a tweezer

d) $2.50/$10.50 = .238, so 23.8% markdown

e) $12.5/.15 = $83.33, $83.33 x .85 = $70.83 selling price

f) .80 x .85 x $ 33.00 = $22.44 a set

g) .90 x 1.65 x .82 x 1.055 x $24.60 = $31.60 cost

Assignment 5-3 Markdowns

Show all work!

1. Body Works sells hand cream at $15.50 a container. A month later, they marked down the hand cream 45% in order to get rid of the inventory. What is the new selling price? __________

2. Galaxy Cosmetics ran a 20% markdown sale off the selling price of their special lotion. The lotion has been selling for $25.00 a bottle. What is the new selling price after the markdown?

   ____

3. The WholeBody Spa sells moisture at $34.50 a tube. After several weeks, they marked down the tube price 28%. What is the new sales price of a tube? __________

4. Jericho bought a massage Table from a wholesaler. He got a 20% discount that amounted to $250 off the list price of the table. What was the original list price?

____

5. BettyBoop's Beauty Shop sells a special bubble bath soap for $19.00 a bottle. After a month, they markdown the remaining bottles 5% to try to get them sold. After another month, the few remaining bottles are discounted 15%. What is the price of these remaining bottles?

6. Elena Elegant Beauty original listed Hair Conditioner at $14.00 a bottle. After a month, Elena took off $3.50 a bottle to try to reduce inventory. What was the % markdown on the bottles?

____

7. Louie's Hair wants to buy shampoo from his favorite distributor, RealWorld Wholesalers. The shampoo is listed at $320.00 a carton. RealWorld offers Louie a 30% trade discount. Louie agrees to buy 3 cartons. How much does Louie send the distributor? ___________

8. Bella Tulio's salon buys fashion mirrors for sale at the salon. The mirrors list for $35.00 a piece from the wholesaler. The wholesaler offers a 10% discount. What is the price to Bella Tulio if the salon purchases ten of the mirrors? ___________

9. Daphne's sells scented candles in her salon. She originally sells them at $22.00 a candle, but after a month she marks them down 20%. After another month she does a second markdown of 15%. Finally, to get rid of the remaining candle, she offers them at a 50% off the price. What is the final selling price of the candles? ______________

10. Guys and Dolls Hair Emporium buys hair conditioner from a beauty supply distributor. The conditioner lists at $10.50 a bottle. The distributor offers Guys and Dolls an 18% discount. Guys and Dolls buys the conditioner, then marks up the bottles 65%. At what amount do Guys and Dolls sell the conditioner? ______________

11. WOW Hair Salon sells hair crème at $12.50 a tube. A month later, they discount the price by 25% . If there is 7% sales tax, what would the customer spend for a discounted tube of crème?

____

12. Sophia's Spa bought massage lotion from a distributor at a list price of $12.50 a bottle. The distributor offers a discount of 10% to Sophia. Sophia marks up the lotion for sale by 90%. After one month, the Spa marks down the remaining inventory 15%. What is the final selling price for a bottle of the lotion? ___________

13. Gertie's Shoppe bought Halloween earrings early in October from the wholesaler at $10.00 each. Gertie marks up the earrings 25%, then when approaching Halloween later in October, Gertie marks up the remaining earrings 20%. Then, after Halloween, Gerties marks down the earrings 50%. What was the final selling price for the earrings? ____________

14. ThelmaLou's, a beauty store, bought CrazySilky Conditioner at $16.50 a bottle from the distributor. ThelmaLou original marked up the bottles for sale 10% and after a month marked them up an additional 15%. After two months they offered the bottle at a markdown of 20% and were able to clean out the inventory. If there was a 6% sales tax, what was the final cost to the customer for a bottle of the conditioner? ___________

15. NancyNails Salon sold sets of fake nails for $22.50 a box, then for the next four weeks marked down the nails by 20% each week. What was the final selling price for a set of nails? ________

16. Wavy Locks Hair offered Haircuts at $25.50 at the beginning of the month. After two weeks they discounted the Haircuts by 22%, then, in the final week of the month , they marked down an additional 11%. If there is a 7% sale tax on this service, what does it cost a Wavy-Lock customer in the final week for a haircut ? ______________

17. Ralph's Tanning Salon was interested in Organic Aloe Vera Gel listed at $19.50 a bottle by HealthBeauty Suppliers. Health-Beauty, as an incentive, offered a discount of 15%. Ralphs bought the gel at the discounted price, and then marked up the bottles by 80%. After a month, the salon marked down the bottles 25%. If there is a 9.5% sales tax, what would a bottle cost one of Ralph's customers?

5-4 SALES, COSTS, AND PROFITS

SALES

Sales are the amount of money a business brings after selling goods or products or after performing services. Sales are also known as **Revenue**.

The total sales for each goods, products, or services is figured by multiplying the price by the quantity sold. So given P = price, Q = quantity, and S = sales, then

$$S = P \times Q$$

For example, Paula's Hair sold 15 bottles of Shampoos at $22.00 a bottle. What was Paula's total sales. P =$22.00, Q= 15, and S = P x Q = ($22.00) (15) = $330.00 or total sales of $330.00.

Example: Ronnie's Beauty Service sold 20 bottles of conditioner at $11.75 a bottle, 41 containers of facial lotion at $5.60 a container, and 15 haircuts at $17.00 a haircut. What was Ronnie's total Sales from these items?

S = ($11.75) (20) + ($5.60) (41) + ($17.00) (15) = $235.00 + $229.60 + $255.00 = $719.60

So Total sales of $719.60.

COSTS

COST is the amount of money a business needs to spend to run its business. This can include the purchasing of goods and products to sell, equipment, and well purchased to operate the physical business, fees charged by local government, and other expenses needed (such as payroll, rent, utilities, and so on.).

Example: KellySue had costs today of:

- 30 bottles of shampoo bought from the distributors at a cost of $12 a bottle
- 25 jars of Hair Gel at a cost of $9.95 a jar
- · $250 for a Humidifier
- 3000 business cards at $50 per 500 cards.
- What is KellySue's total cost?

Shampoo: 30 bottles at $12 per bottle = 30 x $12 = $360.00
Hair Gel: 25 jars at $9.95 per jar = 25 x $9.95 = $248.75 +
Humidifier, 1 x $250 $250.00 +
3000 business card at $50 /500 card , 3000 x $50/500 = $300.00 +

Total cost = $1158.75

PROFIT

Profit is the money that remains for a business after all costs and expenses have been deducted from total sales (or revenues) of goods and services. So Profit dollars = Sales dollars – Cost dollars.

If we let P = profit dollars, S = Sales or revenue dollars, and C = Cost dollars, then

$$\mathbf{P = S - C}$$

It also follows that Sales = Profit + Cost or S = **P + C** and Costs = Sales – Profit or **C = S – P**

Net profit is the difference between the sales revenue and the total cost for the business, which include the cost of the sales plus other expenses, such as rent, overhead, advertising, promotion, etc.)

For example, BettyBoops Beauty Shoppe had total costs for the month of March of \$34,000. The services and products sold to consumers brought in \$39,000 in revenue. To figure BettyBoop's net profit for the month of March, we set cost C = \$34,000 and sales S = \$39,000

And Profit = Sales – Cost, or P = S – C = \$39,000 – \$34,000 = \$5000, net profit.

Example:

In Patty's Beautiful Hair Salon, the profit goal for April is \$6,500. If Patty's total costs and expenses for April was \$9,115, how much in total sales and revenue must Patty create to meet her goal. We make profit needed, P = \$6,500 and cost, C = \$9,115,

And Sales = Profit + Cost or S = P + C = \$6,500 + \$9,115 = \$15,615 in sales needed.

Example:

Perry's Spa had total sales and revenue of \$1,456,000 for the year 2018, and earned a profit of \$365,000. What was Perry's Spa total cost and expenses for 2018?

We make profit, P = \$365,000, and Sales, S = \$1,456,000

And Cost = Sales – Profit or C = S – P = \$1,456,000 = \$365,000 = \$1,091,000 in costs for 2018.

We can use these concepts to do other types of sales, cost, and profit problems.

Example:

Macron purchased seven bottles of conditioner at $8 each for his distributor. Macron wants to make a profit of $12 on each bottle he sells at his shop.

If Macron sells all seven bottles, how much does Macron need to price a bottle and how much in total sales dollars is needed to achieve his goal?

Sales = Profit + Cost or S = P + C.

We calculate total cost as seven bottles at $8 per bottles = 7 x $8 = $56 = C

We can calculate total Profit desired as seven bottles at $12 per bottle = 7 x $12 = $ 84 = P

So S = P + C = $84 + $56 = $140 in sales needed.

And this $140 sales is for seven bottles sold, we figure $\frac{\$140}{7\ bottles}$ = $20 per bottle, the price we need to sell the bottles at.

You can also think of the sales price per bottle in this in this way.

$$\frac{Sales\ \ price}{bottle} \text{ minus } \frac{Cost\ \ price}{bottle} = \frac{profit}{bottle}.$$

As the cost per bottle is $8 and the profit per bottle needed is $12, then

$$\frac{Sales\ \ price}{bottle} - \frac{\$8}{bottle} = \frac{\$12}{bottle} \quad \text{or}$$

$$\frac{Sales\ \ price - \$8}{bottle} = \frac{\$12}{bottle}, \quad \text{so Sales price} - \$8 \text{ must equal } \$12$$

Or Sales price = $8 + $12 = $20 per bottle

And if there are seven bottles sold, S = 7 bottles x $20/bottles = $140 total sales.

Example:

On Tuesday, Kalinda charged $35.00 per manicure done and paid a total of $36.00 for her supplies. These supplies were her only expenses for Tuesday.

After taking out her cost of supplies, Kalinda earned $ 279.00 (profit) for the day.

How many manicures did Kalinda perform on Tuesday?

Sales = Profit + Cost or S = P + C

We make cost, C = $36.00 and Profit, P = $279.00

And S = P + C = $279.00 + $36.00 = $315.00, Kalinda's total sales

As each Manicure was \$35.00, then \$315.00 / \$35.00 = 9 manicures. So Kalinda did 9 manicures in all on Tuesday.

Example:

Marshall wants to offer nail lacquer to his salon customer. He bought a case of nail lacquer bottles from the Beautiful World Beauty distributor. The list price from Beautiful World for the polish was \$18.50 a bottle. The case held fifty bottles in all.

Marshall marks up the bottles 90% in order to make a profit off the sale of the nail lacquer in his salon.

What is Marshall's selling price per bottle, and, if Marshall sells all the bottles, what was his profit?

Profit = Sales – Cost or $P = S - C$

Figuring his cost, C, for fifty bottles at a cost of \$18.50, C = 50 x \$18.50 = \$925.00

Figuring the Selling price per bottle, we must find the markup on the cost. A 90% markup on each bottle cost is (1 + .90) (\$18.50) = (1.90) (\$18.50) = \$35.15 a bottle.

Figuring total sales, S, for fifty bottles at the \$35.15 selling price, S = 50 x \$35.15 = \$1757.50

Figuring Marshall's Profit P, P = S – C = \$1757.50- \$925.00 = \$832.50 profit.

The selling price is \$ 35.15 a bottle, and Marshall earned a profit of \$832.50 if all fifty of the bottle are sold.

Example:

During the month of May, Sheila did eighty full head hair color services. She charged a regular price for the first sixty hair services, then offered a discounted price of 20% off the regular price for the remaining twenty hair services. Her total cost for supplies and expenses for May was \$2100.00.

What was Sheila's regular charge per service and what was the discounted charge per service if she brought in \$3,220.00 in profit for the month of May?

We need total dollar sales for the month. Sales = Profit + Cost, or $S = P + C$
We make Cost, C = \$2100.00, and Profit, P = \$3220.00.
So total Sales, S = P + C = \$3220.00 + \$2100.00 = \$5320.00

To figure the regular charge, make H = the regular hair service charge.
A 20% discount means 1 – .20 = .80 of the original, so the discounted charge = .80 H

And sixty hair services at the H charge is 60 times H or 60H in total sales
And 20 hair services at the .80H charge is 20 times .80H or 20 (.80H) = 16H in total sales.

For all the hair services, total sales = 60H + 16H = 76 H. With Total Sales = S = \$5320.00

Then 76 H = \$5320.00 dividing both sides by 76

Or $$H = \frac{\$5320.00}{76} = \$70.00$$

And .80(H) = .80(\$70) = \$56.00

So to bring in a total Sales Revenue of \$5320.00, she charged sixty hair ervices at \$70.00 a service and twenty hair services at \$56.00 a service

FORMULA TABLE

Real World Problem	Variables Defined	Formula
Tax	P = price; R = tax rate; T = tax	T = P x R
Discount	P = price; R = discount rate; D = discount	D = P x R
Sales Price	P = price; T = tax; D = discount; S = sales price	S = (P-D) + T
Sales	P = price; Q = quantity sold; S = sales	S = P x Q
Cost	P = price; Q= quantity purchased ; C = cost	C = P x Q
Profit	S = sales; C = cost; P = profit	P = S – C

Practice

a) In December, Hank's Hair Salon showed a total revenue for services and sales of $23,818.02. Hank's total costs and expenses for December was $17,159.90. What, if any, was Hank's profit for December?

b) Bella's Beauty Boutique had a profit of $65,300.00 for 2019. If Bella's total sales and revenue for 2019 was $89,300.32, what were Bella's 2019 cost and expenses?

c) At Horatio's Hair, independent hair specialist Sheila did 98 haircuts during April. She incurred out of pocket expenses of $130.00 each week. If she earned $24.50 from each haircut customer, what was Sheila's profit for April (assume four weeks in April)?

d) Daphne purchased 150 bottles of conditioner at $11.30 each for her wholesaler. Daphne wants to sell them at her shop. If she sells all the bottles and shows a profit of $540 from the sales, how much per bottle did she sell the conditioner for?

Answers

a) $23818.02 – $17159.90 = $6658.12 profit.

b) $89300.32 – $65300.00 = $24000.32 costs

c) cost was 4 x $130 =$520.00, Sales were 98 x $24.50 = $2401.00, $2401.00 – $520.00 = s$1881 profit

d) cost was 150 x $11.30 = $1695.00, $1695.00 + $540.00 = $2235.00 in sales. $2235.00 / 150 = $14.90 selling price per bottle.

5-4 Assignments

Show all work

1. In May, Doty's Beautiful Hair Salon showed a total revenue for services and sales of $16717.32. Doty's total costs and expenses for May was $13456.90. What, if any, was Doty's profit for May?

2. Helen's Spa and Treatment had a profit of $56,000.00 for 2018. If Helen's total sales and revenue for 2018 was $98,000.32, what were Helen's 2018 cost and expenses?

3. Debbie's Beauty Boutique had costs for April of $9800, and earned a profit of $3250 in April. What was the total sales and Revenue brought in by Debbie in April?

4. For the first Quarter of the business year, KellySue incurred $11000 in costs and expenses. She earned a profit for the Quarter of $ $18000. What was KellySue's revenue for the first quarter?

5. Working as an independent hair stylist at Fabio's Fabulous Hair, Teri did 21 hair stylings during the first week of January. She incurred out of pocket expenses of $140 for the week. If she earned $35 from each hairstyling customer, what was Teri's profit for the week?

6. Gordo purchased 120 bottles of shampoo at $7.56 each for his distributor. Gordo wants to make a profit of $ 15 on each bottle he sells at his shop. If Gordo sells all 120 bottles, how much does Gordo need to price a bottle and how much in total sales dollars is needed to achieve his goal?

7. Melanie wants to offer hair gel to her salon customers. She bought a case of hair gel jars from Beautiful Body Distributors. The list price from Beautiful Body for the gel was $14.50 a jar. The case held sixty jars in all. Melanie marks up the jars 100% in order to make a profit off the sale of the gel in her salon.
What is Melanie's selling price per jar, and, if Melanie sells all the gel jars, what was her profit on the sale?

8. On Wednesday, Sheila charged $55.00 each for the massages she did on customers. Sheila's total expenses for her supplies was $36.00. These supplies were her only expenses for Wednesday. After taking out her cost of supplies, Sheila earned $ 239.00 (profit) for the day. How many massages did Sheila perform on Wednesday?

9. During July, MissySue did seventy hair stylings in her hair salon. This was her total work load for July. She charged a regular price for the first fifty hair stylings, then offered a discounted price of 30% off the regular price for the remaining twenty hair services. Her total cost for supplies and expenses for May was $1500.00.
What was MissySue's regular charge per service and what was the discounted charge per service if she brought in $2,340.00 in profit for the month of July?

10. Here was a ledger of costs and sales for Yvette's business for the month of April.

cost, April	
Chair Rental	$350
supplies	$560
equipment	$250
taxes	$35
License	$20
Miscellaneous	$55

Revenue, April
40 Haircuts @$18 per haircut
25 Pedicures @ $20 per pedicure
15 Facials @ $22 per facials
18 Hair coloring @ $35 per hair coloring
15 manicures @ $23 per manicure
$310 in tips

What was Yvette's profit for April? __________

5-5 REPRESENTING DATA

When desiring a certain hairstyle, hair color, or a makeup style, a customer may come into your business with a photo to illustrate the desired look. The familiar saying "A picture is worth a thousand words" certainly holds true, as it can give you a far better understanding of what the customer wants than having he or she explain it to you in words.

When working with numerical data in one's business, being able to present that data in "pictures" helps to demonstrate patterns or trends in sales, costs, and profits and to build an understanding of financial position. Math "picture" creation involves taking data and creating graphs. Becoming proficient and confident in the creation and use of data and graphs is important as one build his or her professional business.

There are many types of graphs that can be used. We will explore a few of them here.

GRAPHS

What is a graph? In math, a graph can be defined as a pictorial representation or a diagram that represents data or values in an organized manner. A well-organized graph will illustrate data so that the viewer can easily compare and establish patterns regarding numbers and outcomes over a period of times or other change periods.

DATA

Graph creation starts with data. Data is a collection of quantitative values. Examples of data would be

- The number of haircuts, facials, and pedicures done in April
- the list of cost expenses for January
- All the sales totals in a day
- The markup number of a sale
- The profit earned each month

And so on . . .

This numerical can be collected in placed in different categories and organized in tables or charts. And then this data can be displayed in a graph.

BAR GRAPHS

One of the oldest and most common ways to represent numerical data is the BAR GRAPH. A Bar graph has two axes, one to measure a specific category of events and the other the numerical frequency of the event happening.

Example: Bosco collects data on how many Reality TV shows appear on the cable lineup each day.

Here are the numerical data facts:

day	number of reality shows
Monday	15
Tuesday	16
Wednesday	26
Thursday	21
Friday	12
Saturday	31
Sunday	26

One axis will be the category – day of the week, and the other axis will be the numerical data, the number of reality shows on each day.

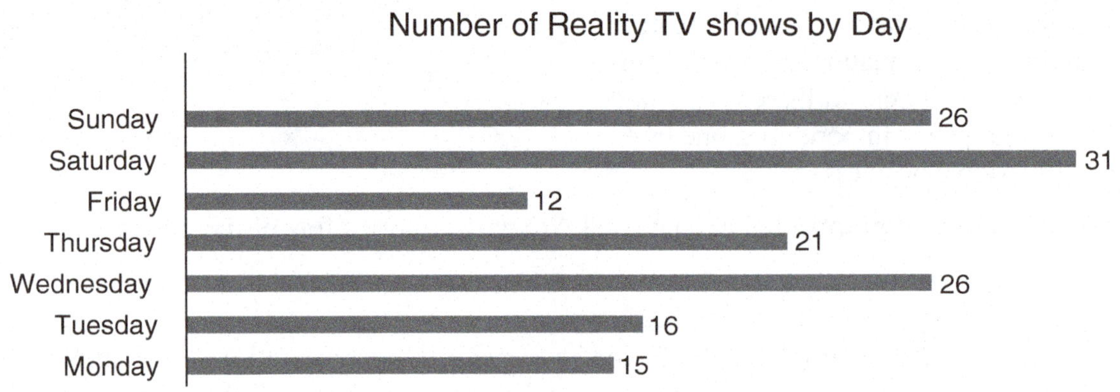

Or

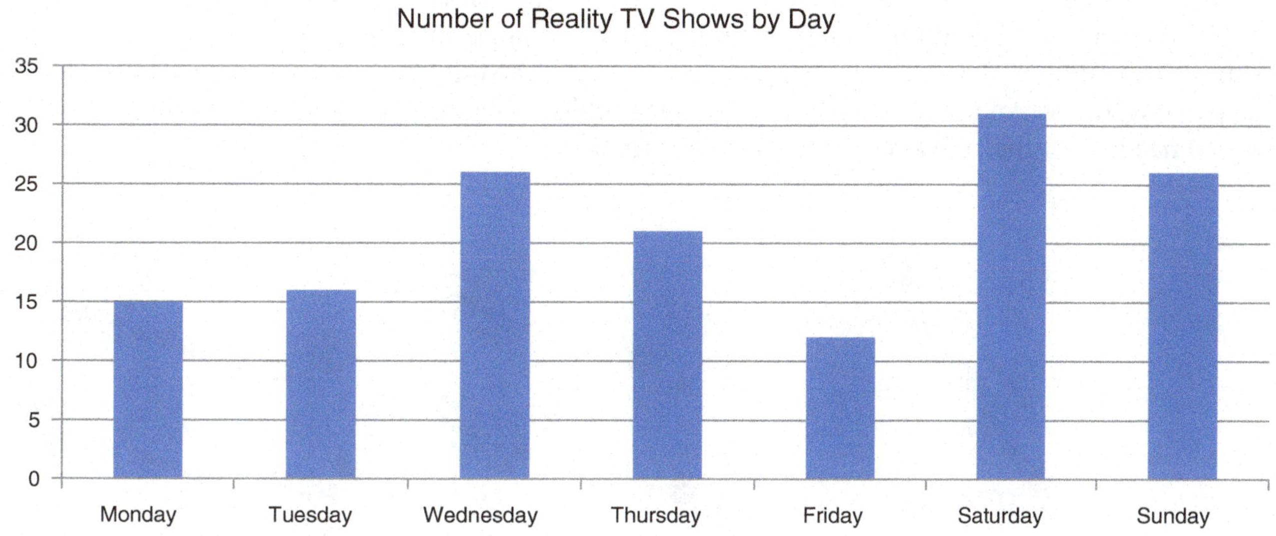

By showing data in this way, one can easily interpret which day has the most Reality TV, (Saturday) which day has the least (Friday) and so on.

Some types of Bar Graphs have two bars for each category – called double-bar graphs, some have three bars (three-bar graphs,) four bars, etc. These can break down original categories into sub-categories, for comparison. Here is a double-bar graph example:

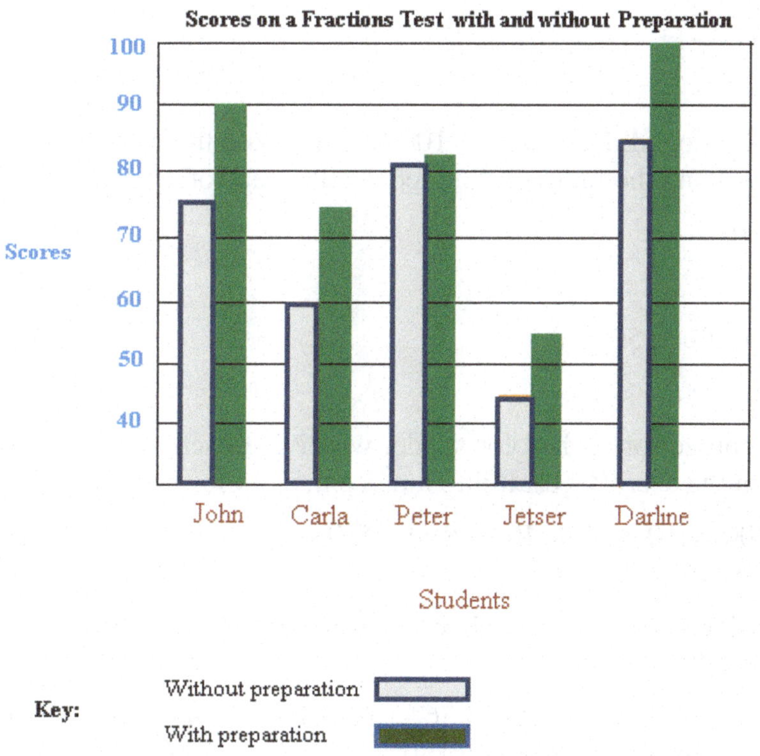

PIE GRAPHS (or CIRCLE GRAPHS)

Another popular graph for representing data is the pie graph. It is also sometimes known as a circle graph. This graph is based on a circle with "pie slices" representing the percent of some event happening. The number of some event happening divided by the total possible number of events creates the percent fraction of the whole. This fraction is multiplied by 360° in order to get the central angle degree needed for the "pie slice". Here is an example.

Greta's GreatBeauty Store did a survey of 100 customers on the preference of the red shade color of a new nail polish. Among the four possible red shades, the voting came out

- Radical Red : 10
- Rose Red : 25
- Cherry Blossom Red : 50
- Spice Red: 15

To figure the percentage for each color shade, we divide each vote total by the whole number of votes, 100, and turn the decimal result into a percent, as follows

Radical Red, 10/100 = .10 = 10%; Rose Red, 25/100 = .25 = 25%; CherryBlossom Red, 50/100 = 50%; Spice Red, 15/100 = .15 = 15%.

We now need the degree of the "pie slice" for our pie graph. We figure this by multiplying the percent by 360°, as follows

Radical Red, 10% x 360° = .10 x 360°= 36° ; Rose Red, 25% x 360° = .25 x 360°= 90° ; Cherry-Blossom Red, 50% x 360° = .50 x 360°= 180° ; Spice Red, 15% x 360° = .15 x 360°= 54°

RED NAIL POLISH SHADE PREFERENCE

color	total responses	fraction of total	%	degree of pie slice
Radical Red	10	10/100	10%	36°
Rose Red	25	25/100	25%	90°
CherryBlossom Red	50	50/100	50%	180°
Spice Red	15	15/100	15%	54°

We put this data in a pie graph

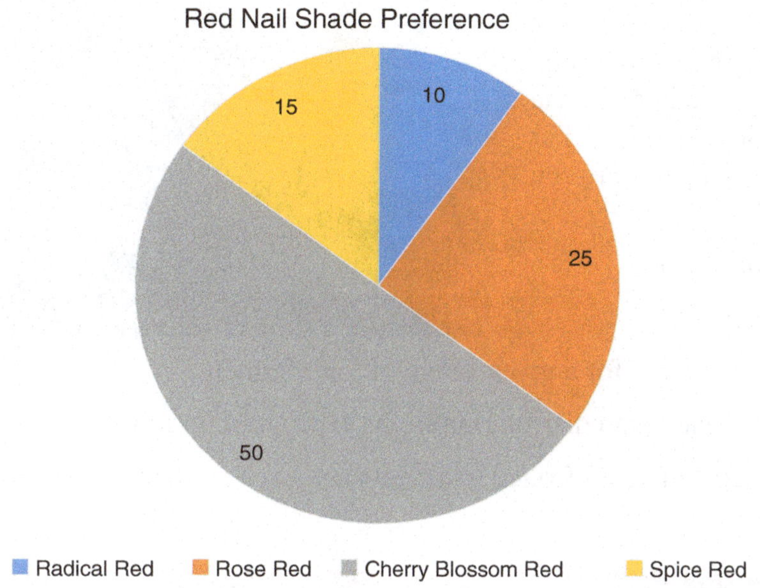

And here are details for each "pie slice."

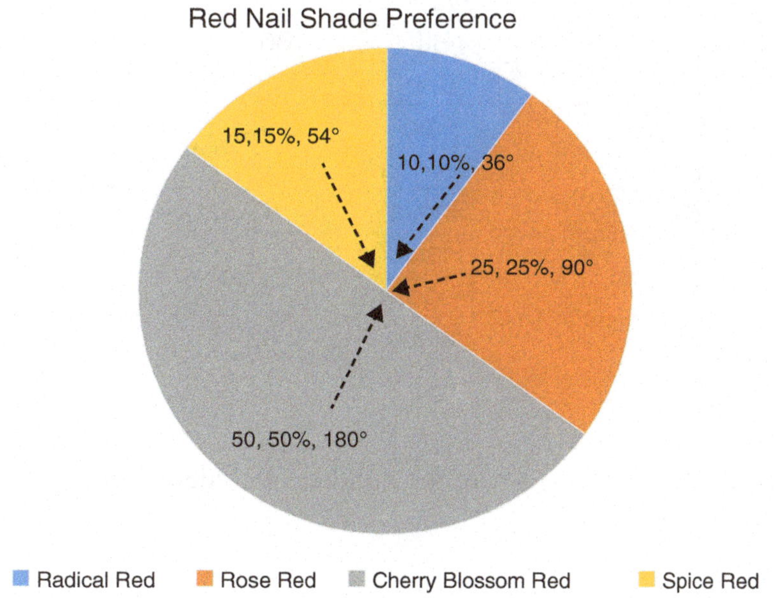

A circle graph presents a visual representation of the numerical data by percent to whole, which allows for quick comparison among categories. Based on the "pie slice", one can easily see from the graph above that CherryBlossom Red is the preferred choice by a big margin; Radical Red is the least preferred, and so on.

HISTOGRAMS

When data is spread over a wide range with many different values, a histogram is a smart way to represent the data. It uses a "bar graph" approach, but the "categories "are a range of values along one axis and the frequency inside that range is the other axis. In addition, the bars sit side by side, unlike a bar graph where there is a gap between categories.

To create a histogram, you

- collect data, then
- create a frequency table for a given range of values, then
- we put this frequency data distribution in the Histogram graph.

Example: Here is data collected from 31 students and each student's average time completion a hair coloring.

student	minutes		student	minutes		student	minutes
Bill	67.5		Quinn	77.5		Shelley	84
Sandy	78		Sydney	78		Trenton	83.5
George	67.5		Biff	86.5		Horatio	71
Ted	63		Eric	69		Pat	76
Greta	84		Fred	74.5		Kathi	78
Delores	71		Gordon	71		Boris	74
Wendall	79		Sheila	63		Louie	89
Ronnie	61		Hank	71		Nelson	75.5
Horatio	76		Reggie	76.5		Orel	82.5
SallyJo	72		Valerie	83		Jake	79
Bosco	71						

Turning this into a frequency table, where the number of student's times within 5 minutes, starting with 60 minutes, would look like

FREQUENCY OF HAIR COLORING TIMES

minute time range	Number
60-65	3
65-70	3
70-75	8
75-80	10
80-85	5
85-90	2

And the DATA in a histogram looks like

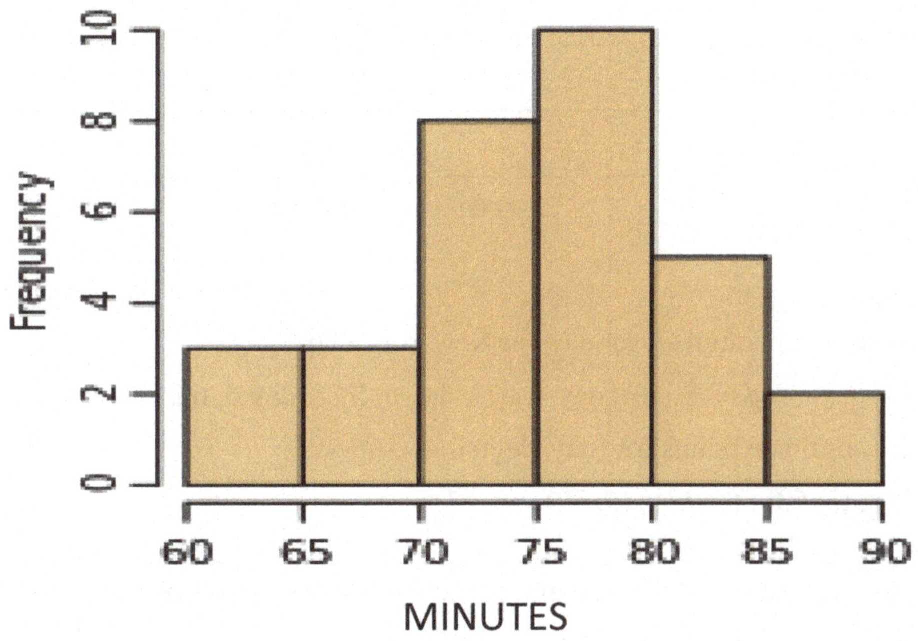

The value of a histogram is it provides a pictorial representation of the distribution flow of collected data. For instance, the graph above shows the majority of the average minutes' data is distributed over the 70 to 80-minute range. This demonstrates that the data is heavily distributed over the middle range of the minutes.

LINE GRAPHS

A line graph is created by a sequence of points created by data values, connected by segments Line graphs often are used to show how data changes over times. We can use a two-coordinate graph method to graph the data.

Example: data was collected measuring the population growth each decade of the United States from 1960 to 2000 as follows

year	1960	1970	1980	1990	2000
Total	9,579,677	11,198,655	14,229,191	16,986,510	20,851,820

We have five 2 coordinate points for (year, population) as follows

(1960, 9579677) , (1970, 11198655) , (1980, 14229191) , (1990, 16985510) , (2000, 20851820)

With the horizontal axis representing the year values and the vertical axis representing the total population, we graph the points and connect each point with a segment as follows

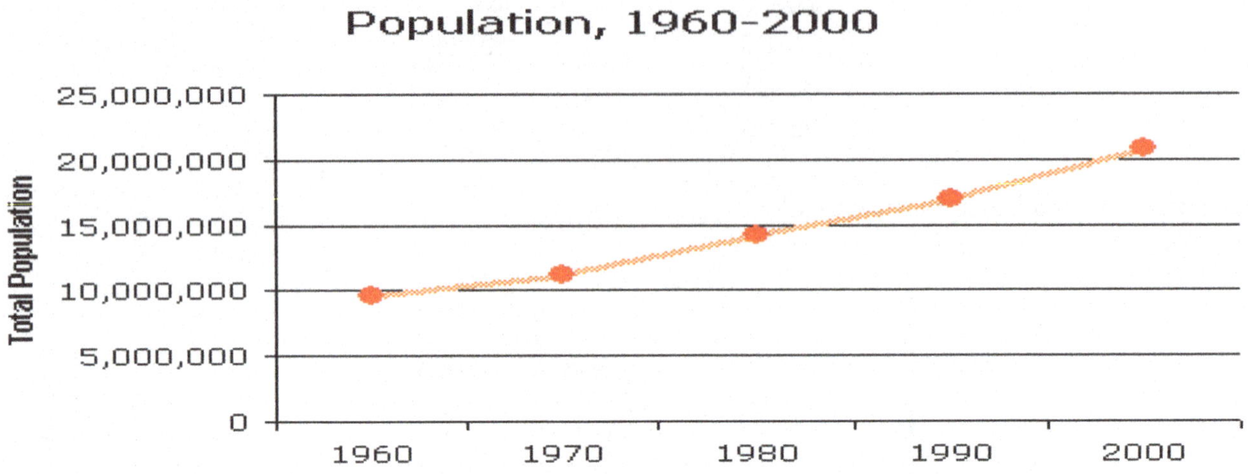

Another Example:

Collected data for temperatures over days in New York City

Day 1, 43°; day 2 , 53°; day 3 , 50°; day 4, 57°; day 5, 59 °; day 6, 67°

We have six 2 coordinate points for (day, degree) as follows

(1, 43) , (2, 53) , (3, 50) , (4, 57) , (5, 59) , (6, 67)

With the horizontal axis representing the day values and the vertical axis representing the degree of temperature, we graph the points and connect each point with a segment as follows

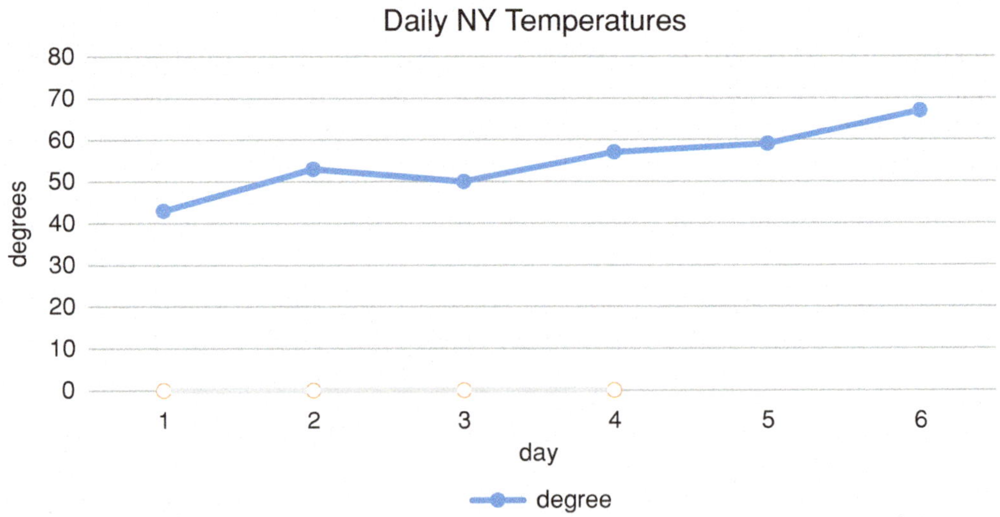

You can see a trend of the data over time in both of these graphs.

SALES, COST, AND PROFIT FORECASTING

Creating line graphs for sales dollars and cost dollars overtime on the same coordinate graph can be used to do forecasting future sales and cost trends and future profit possibilities.

Where the line graph for sales intersects the sales line graph is where sales and costs are equal and the point of intersection is called the BREAK-EVEN point. The break-even point is critical to determine for a business, as it will tell what sales are needed to turn a positive profit flow, the definite place your business wants to be to be successful.

Let's look at an example.

You prepare and sell items in your store. In the course of the business day, your average costs amount to this

- one time cost at the start of the day of $110.00
- Every hour thereafter a $10 cost

  Your average sales received are

- $40 per hour.

Data created for costs per hour in a table is

hours	0	1	2	3	4	5	6	7	8	9	10
$ cost	120	130	140	150	160	170	180	190	200	210	220

We have cost data points for (hour, $ cost) of

(0,120) , (1,130) , (2,140), (3,150), (4,160), (5,170) , (6,180), (7,190), (8, 200), (9,210), and (10,220)

With Hours the horizontal axis and $ cost the vertical axis, a line graph for this data would be

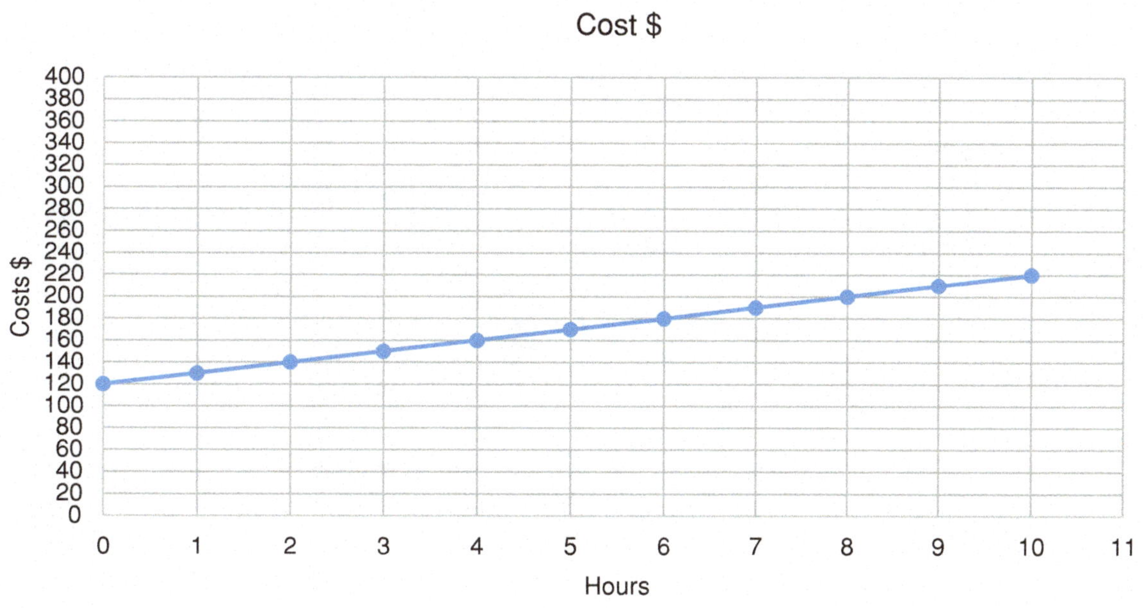

Data created for sales per hour is

hour	0	1	2	3	4	5	6	7	8	9	10
$ sales	0	40	80	120	160	200	240	280	320	360	400

We have sales data points for (hour, $ cost) of

(0,0) , (1,40) , (2,80), (3,120), (4,160), (5,200), (6,240), (7,280), (8, 320), (9,360), and (10,400)

With Hours the horizontal axis and $ sales the vertical axis, a line graph for this data would be

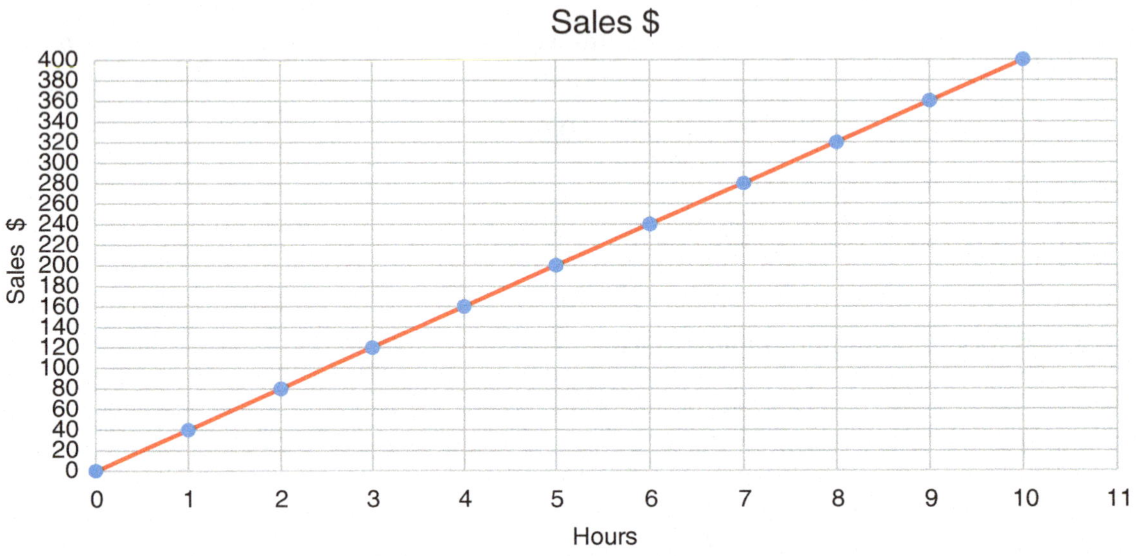

If we put the cost line graph and the sale line graph on the **same** coordinate graph, we get

Breakeven point (4,$160)

You will notice that the line graph intersects at the point (4,$160) which represents 4 hours and $160. This tells us that after 4 hours of business, sales and costs are the same.

In business, this point is referred to as the **BREAKEVEN** point. When reaching the breakeven point for costs and sales, a business knows if sales and costs continue in their same respective pattern with the time period, for any number of time periods above the breakeven point, total sales dollars will exceed cost dollars and a positive profit occurs. Conversely, if a business is operating below the number of time periods on the breakeven point, costs are exceeding sales, and there is a negative profit.

For our example, sales in the 5th hour are $200 and costs in the 5th hour are $170, so sales exceed costs by $30, so a positive profit. Sales in the 6th hour, is $240 and cost is $180, so sales exceed costs by $60. The amount of profit gain in this situation grows at an increased manner the farther we get from the breakeven point. In your business, you want sales to exceed costs and to achieve a growing profit, so the breakeven point is important to recognize.

Example:

DonnaMae in her beauty shop generated revenue for 2019 as follows:

HEALTH SERVICES OR PRODUCTS	Massages	Hair styling, Haircuts, Hair coloring	Pedicure and Manicures	Facials	Retail sales
2019 Revenue	$25,000	$55,000	$15,000	$30,000	$20,000

DonnaMae calculates that her shop has 2500 client appointments.

She wants to find the average price per service or product in 2019 so she can plan for revenue in 2020. To find this, Donna adds all the total revenue

$25000 + $55000 + $15000 + $30000 + $20000 = $145000 total revenue.

She then divides this by the total number of appointment client appointments.

$145000 ÷ 2500 = $58.00. Therefore, the average revenue per client appointment was $58.00 in 2020.

Donna plans to increase the number of client appointments in 2020, and wants to use 2019 data to analyze potential revenue.

Let's put this in a line graph for Revenues versus a number of client appointments.

As the revenues equal the number of client appointments times $58.00. Here is a table of data

Number of client appointments	0	500	1000	1500	2000	2500	3000	3500	4000
Number of client appointments x Revenue $ per client appointment	0 x $58 = **$0**	500 x $58 = **$29000**	10000 x $58 = **$58,000**	1500 x $58 = **$87,000**	2000 x $58 = **$116,000**	2500 x $58 = **$145,000**	3000 x $58 = **$174,000**	3500 x $58 = **$203,000**	4000 x $58 = **$232,000**

Making our axes Revenue $ and Number of Client Appointments, the graph is

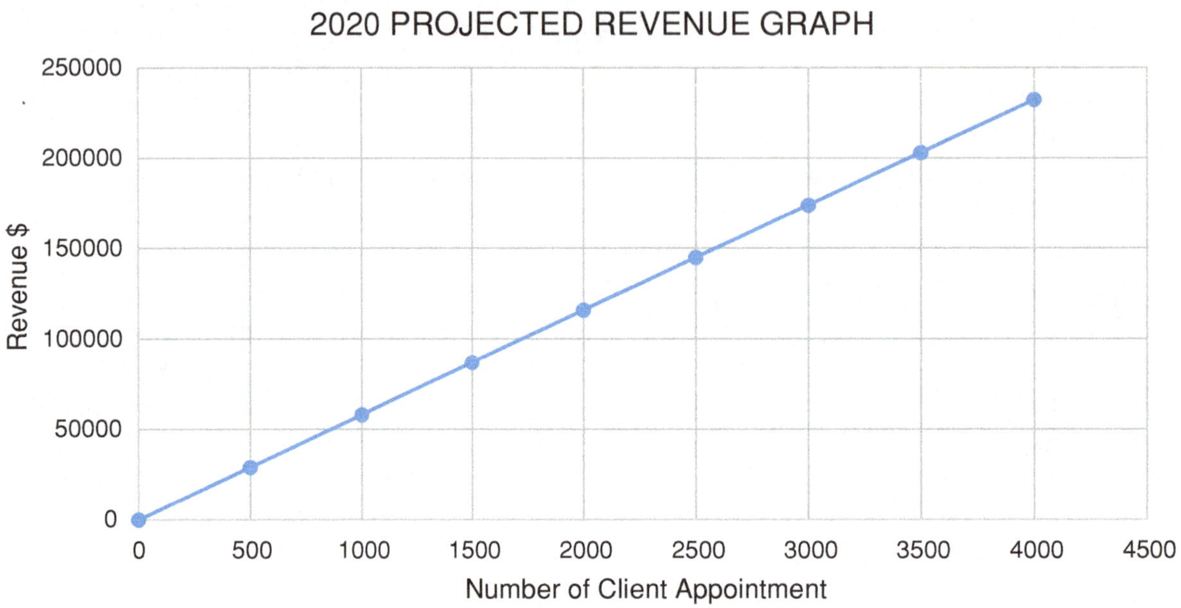

DonnaMae also calculated her costs for 2019 as follows

Fixed costs – those costs that are independent of the number of clients

fixed cost type	Rent	utilities	taxes	licenses	advertising	miscellaneous
Fixed cost $	**31000**	**2000**	**1500**	**200**	**3300**	**1000**

Variable costs – those costs impacted by the number of clients

Variable cost types	labor	store products	supplies	miscellaneous
Variable cost $	**18000**	**12000**	**16000**	**1500**

Figuring the total Fixed costs for 2019, DonnaMae added

$$\$31000 + \$2000 + \$1500 + \$200 + \$3300 + \$1000 = \$39,000$$

To figure the $ cost per client, DonnaMae added all the variable costs then divided by the 2500 client appointments

$18,000 + 12,000 + 16,000 + 1500 = $47,500, $47,500÷ 2500 = $19.00 per client appointment.

DonnaMae now wants to use 2019 data to analyze potential costs. Let's put this in a line graph for total Costs vs number of client appointments.

Total costs equal fixed costs of $39000 + number of client appointments times $19.00. She records

Number of client appointments	0	500	1000	1500	2000	2500	3000	3500	4000
Total Fixed costs$ + client appointments x Variable Costs$ per client appointment	39,000 + 0 x $19.00 = **$39,000**	39,000 + 500 x $19.00 = **$48,500**	39,000 + 1000 x $19.00 = **$58,000**	39,000 + 1500 x $19.20 = **$67,500**	39,000 + 2000 x $19.00 = **$77,000**	39,000 + 2500 x $19.00 = **$86,500**	39,000 + 3000 x $19.00 = **$96,000**	39,000 + 3500 x $19.00 = **$1,05,500**	39,000 + 4000 x $19.00 = **$1,15,000**

Making our axes cost $ and number of client appointments, the graph of this would look like

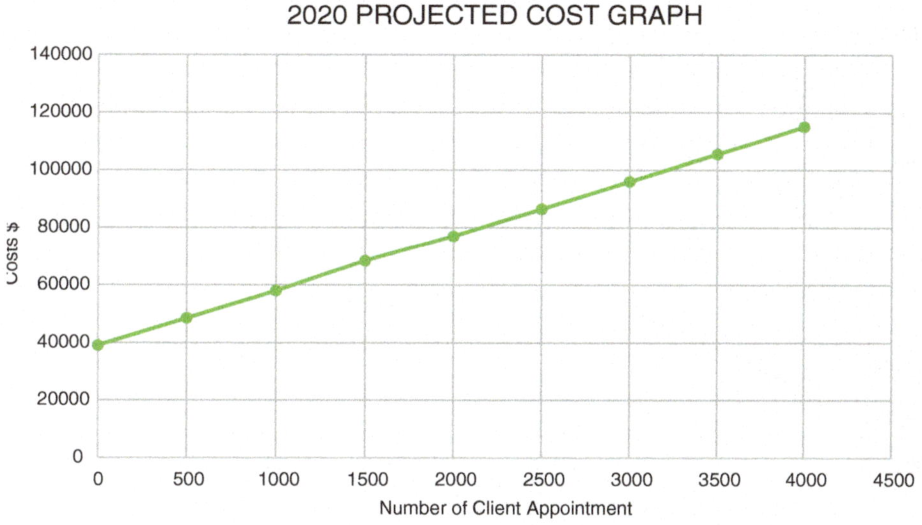

DonnaMae wants to figure the Break Even point for projected 2020 sales and costs. She puts the Sales and Costs data on the same coordinate graph, with axes Sales or Costs $ and Number of client appointments.

Breakeven point (1000, $58000)

The two line graphs intersect at point (1000, $58000), so the breakeven point is 1000 client appointments and $58,000 in sales or costs. Reaching and then exceeding 1000 client appointments in 2020 should result in more revenue than cost and therefore a profit.

DonnaMae now wants to figure and represent the amount of profit given the number of client appointments. Taking the sales and cost data, she records the data

Number of client appointments	0	500	1000	1500	2000	2500	3000	3500	4000
Sales – Cost = Profit	$0 –$39,000 = **–$39,000**	$29,000 – $48,500 = **–$19,500**	$58,000 – $58,000 **= $0**	$87,000 – $67,500 = **$19500**	$116,000 – $77,000= **$39,000**	$145,000 – $86,500 = **$58,500**	$174,000 – $96,000 = **$78,000**	$203,000 – $105,500 = **$97,500**	$232,000 – $115,000 = **$117,000**

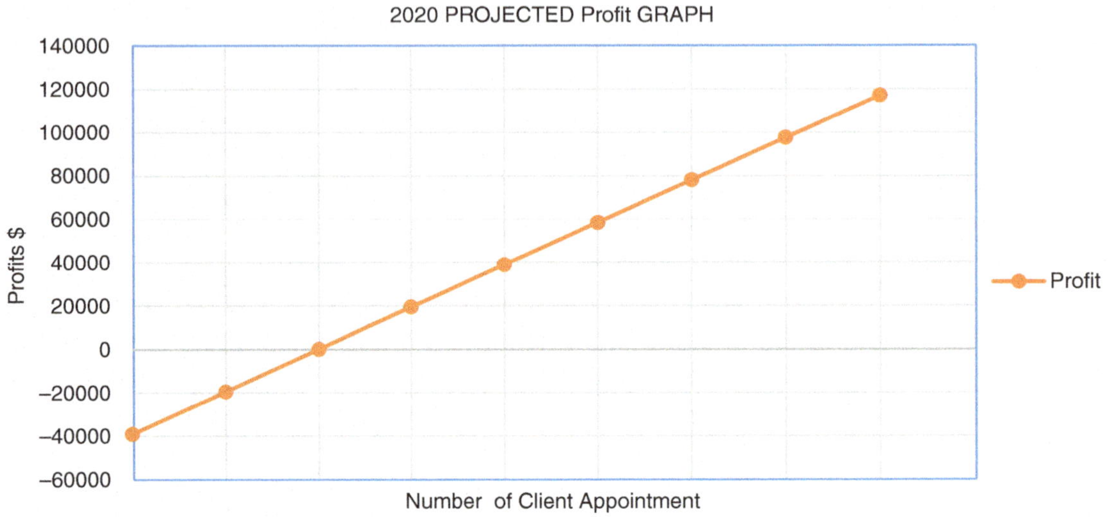

By this graph, Donna sees that not until she reaches 1000 client appointments will she sees a positive profit.

5-5 ASSIGNMENTS

1 and 2. Here is data for sources of public school in Federal, State, and local taxes for 6 states. Use it for questions 1 and 2

state	federal tax %	State tax %	local tax %
Hawaii	8	90	2
Vermont	2	86	12
Minnesota	7	73	20
Kansas	20	50	30
Massachusetts	5	47	48
Washington	8	62	30

1. Create a bar graph showing the **% Federal Tax** for the six states
2. Create a pie chart for the three sources of revenue for **Kansas** to the nearest whole number %

3. Data was recorded for the time in seconds that phone-in customers were on hold at Bella's Twilight Salon during the day.
Here is the data

9.9	6.4	5.6	3.2	4.5	5.2
12.2	4.4	10.3	11.5	5.5	4.4
5.6	2.4	14.5	4.6	14.8	8
8.8	6.6	9.2	13.3	7.7	3.2
6.7	5.5	7.4	18.2	4.1	8.6
4.4	5.5	6	11.5	6.8	11.2

With the time in seconds intervals of 2.5 seconds, starting with 0 seconds, create a frequency table, and a histogram.

4. This pie graph represents data for favored preference of hair products from a voting poll of 150 customers.

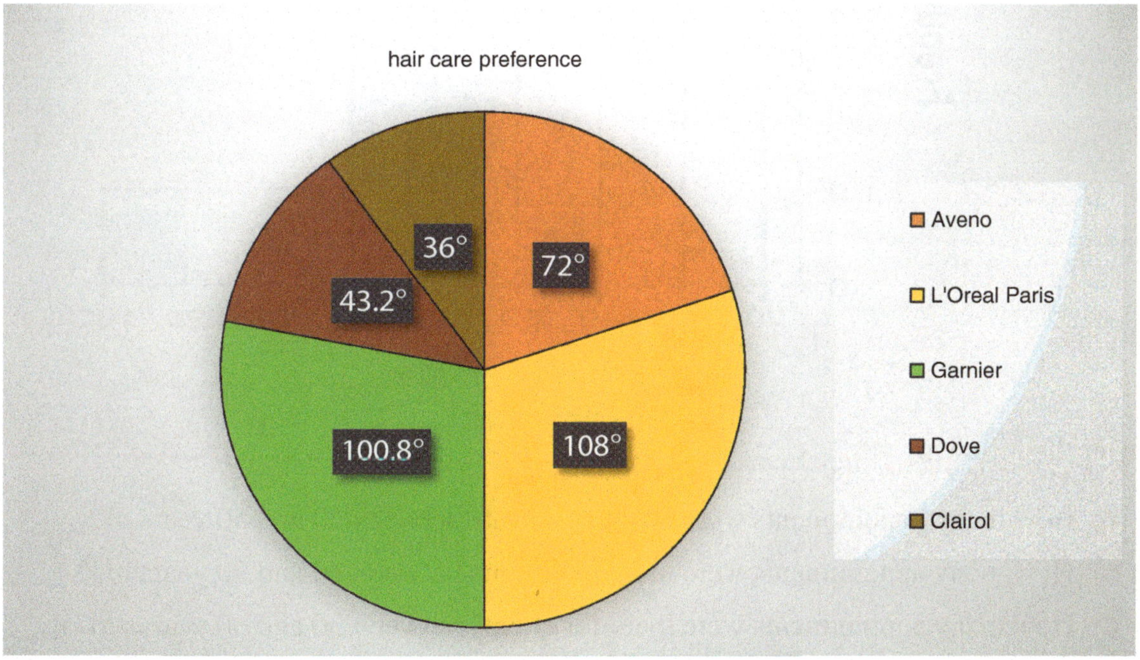

Given the degree of the pie slice angle, determine the number of votes and % of the 150 customers each hair care product received.

Hair product	Number of votes	% of vote
Aveno	________	________
L'Oreal Paris	________	________
Garnier	________	________
Dove	________	________
Clairol	________	________

5. The following data should the number of customers seen for the first 8 weeks at Harry's Hair Emporium.

week	1	2	3	4	5	6	7	8
customers	28	32	26	15	20	22	23	28

Create a line graph to represent this data.

6. The following is a histogram representing the frequency of appointments for a week at a beauty spa, by age of client.

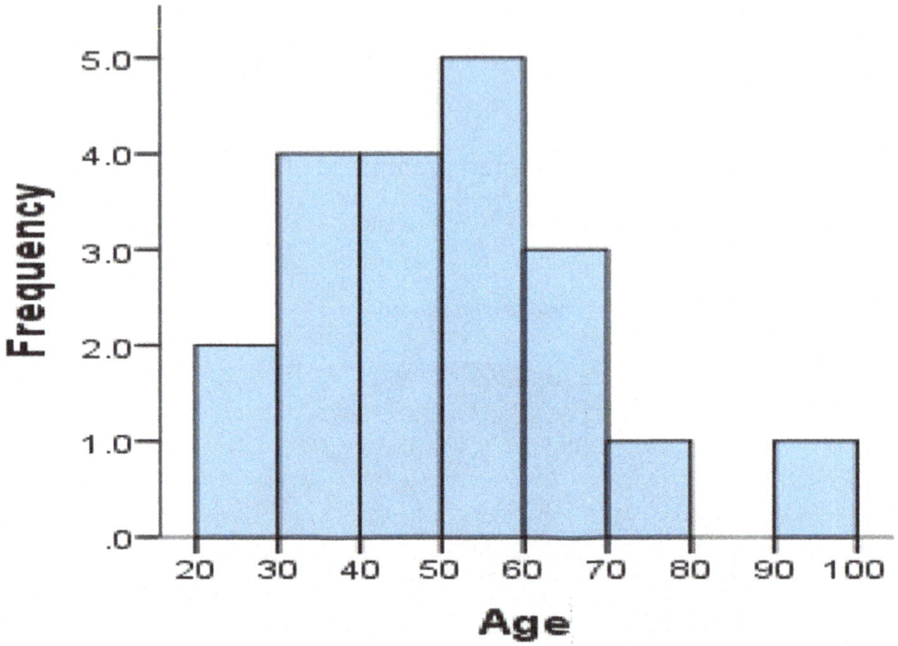

Answer the following

a) How many appointments were there for clients between 40 and 50 years old ________

b) How many appointments were there for clients between 80 and 90 years old? _______

c) How many appointments were there for clients between 30 and 60 years old? _______

d) How many appointments total were there for the week? ______

7. A business found the following occurred in a day

Average costs

- One-time cost at the start of the day of $120.00
- Every hour thereafter a $50 cost

Average sales/revenues

- $70 per hour.

a) Based on this, create tables of data for costs per hour and sales per hour

b) Graph both costs and sales on the same graph with dollars for the vertical axis and hours for the horizontal axis

c) find the breakeven point – the hour where costs and sales are the same.

8. InnerBeauty Spa and Salon generated revenue for 2019 as follows:

HEALTH SERVICES OR PRODUCTS	Massages	Body Treatments	Pedicure and Manicures	Facials	Retail sales
2019 Revenue	$45,000	$35,000	$15,000	$60,000	$20000

InnerBeauty also calculated costs for 2019 as follows

Fixed costs – those costs that are independent of number of clients

fixed cost type	Rent	utilities	taxes	licenses	advertising	miscellaneous
Fixed cost $	25000	5000	2500	2200	3300	2000

Variable costs – those costs impacted by the number of clients

Variable cost types	labor	store products	supplies	miscellaneous
Variable cost $	58000	22000	19000	6000

InnerBeauty has calculated 3500 clients were served in 2019.

They want to use this data to prepare for 2020 sales and costs and profits.

Using the data, Find the

- the average sales/revenue per client
- the average variable cost per client
- Using the average sales/revenue per client, create a table of sales per client (0 – 5000 clients)
- Using the fixed costs and average variable cost per client, Create a table of total cost per client (0-5000 clients)
- Graph both the cost data and sales data on the same graph
- Find the breakeven points, where the number of clients has the same sales and cost
- Create a table for 2019 profit
- Graph the 2019 profit data per client (0 – 5000 clients)

CHAPTER 6

Personal and Business Financing

There is not a profession out there that doesn't deal with MONEY!!!! The same is true for your Cosmetology/Esthetics Career. The better we understand the mathematics of Finance, the better we will be in any of our future cosmetology or aesthetics business work. And, the better we will be in our financial future. As business managers, you may be faced with the need to borrow money to purchase equipment, supplies, and/or real estate and to meet expenses. Also, planning for future business and personal financial goals may mean investing money. We will introduce the concept of interest, both when it is a charge when borrowing money and when it is a growing additive to the balance of a money investment. In both cases, interest acts in very much the same way.

SIMPLE INTEREST

When we invest money or borrow money, we need to be familiar with interest. We will deal with **SIMPLE INTEREST** to start.

We need to define some finance terms.

PRINCIPAL is the money we start with. This could be a lump sum of dollars we want to invest or it could be money we want to borrow. In formulas, we usually use **P** to represent Principal.

INTEREST RATE is the % interest is applied for a given time period. In formulas, we usually use "***r***" to represent the interest rate.

Note: The time period unit needs to match the time period the interest rate applies to. So if r is an annual interest, then t must be in number of years. If r is monthly, t must be in number of months.

TIME is the number of time periods that the investment or loan is carried out for. In formulas, we usually use "***t***" to represent the number of time periods.

INTEREST is the amount of interest that the % interest rate applied to the Principal over the number of time period generates. In formulas, we usually use "***I***" to represent the interest.

END BALANCE is the amount of money is in the investment or is due if a loan after the t number of time units. We can use "***A***" to represent this end balance.

6-1 ONE-TIME SIMPLE INTEREST

Example: Felicity borrowed $100.00 from a bank at a 5% interest rate. This means that Felicity needs to pay back the $100 plus 5% of the $100.00. As 5% of $100.00 = .05 (100) = $5.00, Felicity will pay back the bank $100.00 + .05 (100.00) = $100.00 + $5.00 = $105.00

As she starts with $100.00, that is the Principal, so let P = $100.00. As the interest rate is 5% = .05, we can let r = .05. As the interest paid is $5.00, we can let I = $5.00. As the end balance due of the loan is $105.00, we can let A = $105.00

As simple Interest is figured by multiplying the Principal by the interest rate, we can write this in a variable formula as

$$\mathbf{I = P\,r}$$

As the interest applied is $5.00, we can write an equation as $105.00 = $100.00 + $5.00. Also the interest can be figure as $5.00 = $105.00 – $100.00

Which would make a variable formula of

$$\mathbf{A = P + I \text{ and } I = A - P}$$

In addition, if we use our interest rate in an equation, we can write this situation in an equation as

$$\$105.00 = \$100.00 + (.05)(\$100.00)$$

We can factor out a $100.00 from the two terms on the right side and manipulate this into

$105 = $100.00 (1 + .05). Which would make a variable formula of

$$\mathbf{A = P\,(1 + r)}$$

This formula can be used to find one-time simple interest problems if we know the Principal and the short term interest rate.

Example:

Dannie borrows \$5000 from Beauty "R" US Credit Union for her beauty business at a simple interest rate of 7.55%. How much does Dannie need to pay to clear her loan? And how much was the interest.

P = \$5000 r = .0755, so A = P (1 + r) = \$5000 (1 + .0755) = \$5000 (1.0755) = \$5377.50 that needs to be paid back to the Beauty "R" US Credit Union. And The Interest I = P r = \$5000 (.0755) = \$377.50. As A = P + I, We can also figure this by figuring the End Balance minus the Principal, or I = A − P = \$5377.50 − \$5000 = \$377.50

Example:

Gregory borrowed \$390 from a friend at 11% interest. How much will Gregory need to pay back? And what was the interest paid?

P = \$390, r = .11 so A = \$390 (1 + .11) = \$390 (1.11) = \$432.90 will be paid back to the friend. The interest I = \$390 (.11) = \$42.90 (or figure as \$432.90 − \$390 = \$42.90)

Practice

Find the amount A for the following one-time simple interest scenarios

a) Borrowed \$23500 at 6.6% interest, End Balance __________ Interest __________

b) Borrowed \$1235000 at 4.52% interest End Balance owed __________ Interest __________

c) Borrowed \$62.90 at 2% interest End Balance owed __________ Interest __________

Answers:

a) end balance = \$23500 (1 + .066) = \$25051, interest = \$23500 (.066) = \$1551

b) end balance \$1235000 (1 + .0452) = \$1290822, interest = \$1235000 (.0452) = \$55822

c) end balance = \$62.90 (1 + .02) = \$64.16, interest = \$62.90 (.02) = \$1.26

6-1 Assignments

Show all work

1. A loan of \$45600 accumulated \$555 in interest. What was the ending balance payout on this loan? __________

2. A loan taken out for \$11555 was paid out later at a payout of \$12355. What was the interest incurred on this loan? __________

3. If a payout on a loan was \$678.99 and the interest incurred was \$98.99, what was the principal value of the loan? __________

4. On a loan of \$3520.10, if the one-time simple interest was at 7.77% of the principal, how much interest was incurred? ___________

5. For each one-time simple interest loans, find the ending balance and interest paid.

   a) Borrowed \$980 at 7.6% interest, End Balance ___________ Interest ___________

   b) Borrowed \$1115 at 4.52% interest End Balance owed ___________ Interest ___________

   c) Borrowed \$88.90 at 3% interest End Balance owed ___________ Interest ___________

   d) Borrowed \$33333 at 26.66% interest End Balance owed ___________ Interest ___________

   e) Borrowed \$ 435.22 at 18.55% interest End Balance owed ___________ Interest ___________

6. TerrificSpa needed some new equipment. The owner took out a \$6000 loan to help purchase the equipment. It was a one-time simple loan at 10.5% interest. What should TerrificSpa pay back to clear the loan? ___________

7. Gerrie borrows \$600 from a friend in a one-time simple interest loan at 12.5%. How much does Gerrie pay back to clear the loan? ___________

8. Julie borrowed \$5000 from a lender to use for her hair salon remodeling. It was a one-time simple interest loan at 6.7% interest. What did she eventually pay back to clear the loan? ___________

9. Ronald borrowed \$750 from a friend and paid \$794 back. What was the dollar amount of the interest? ___________

10. Kurt took out a one-time simple loan for \$670. If the interest is 2.22%, how much interest does Kurt pay to clear the loan? ___________

11. The Beauty Store took out a \$3500 store improvement loan as a one-time simple loan at 15.55% interest. What was the interest incurred? ___________
    How much did the Beauty Store pay back? ___________

12. Nelson borrowed \$22500 from a lender. The lender charged a 8.5% one-time simple interest. What does Nelson pay the lender to pay off the loan? ___________

6-2 SIMPLE INTEREST OVER TIME

One-time Simple interest is usually used for short term/time loans. Many simple loans (and investment) are carried over specific time period, like for a month, three months, half a year, one year (annually), many years, etc.

We generalize the simple interest formula as: given **r** the % rate over a time period, and **t** the number of those time periods during the loan or investment, then

$$\mathbf{A = P\,(1 + r\,t)}$$

And the amount of interest paid is

$$I = P\,r\,t$$

The Interest can also be found using the formula as before

$$\text{As } A = P + I \text{ then } I = A - P$$

Example:

Glenda borrowed $2000 from a lender to apply toward her spa business. She did so in a simple interest loan at 7% annual interest for 4 years,

How much does Glenda need to pay at the end of 4 years. And how much interest is paid.

The Principal is $2000, the annual interest rate is 7% or .07, and the Time is for 4 years, so

Let P = $2000, r = .07, and t = 4 and to find the end balance A,

$A = P(1 + r\,t) = \$2000\,(1 + .07 \cdot 4) = \$2000\,(1.28) = \$2560$ to be paid back

The amount of interest can be figured as $\mathbf{I} = P\,r\,t = \$2000\,(.07)\,(4) = \$560$ interest

Or $I = A - P = \$2560 - \$2000 = \$560$

Example:

Horatio Hair business invested $6000 in a 9.2% simple annual interest investment for 9.5 years. How much would Horatio Hair have after 9.5 years and how much interest would he earn.

$P = \$6000$, $r = .092$, $t = 9.5$ then $A = P(1 + r\,t) = \$6000\,(1 + .092 \cdot 9.5) = \$11244$ in the investment balance, and interest earned is $\mathbf{I} = P\,r\,t = \$6000\,(.092)\,(9.5) = \$5244$ or $I = A - P = \$11244 - \$6000 = \$5244$

Example:

PollyAnna borrowed $500 from a colleague at 7.1% simple annual interest and paid it back after 8 months. How much does PollyAnna payback to the colleague and of that, how much interest does PollyAnna have to pay?

$P = \$500$ $r = .071$ for time, as there are 12 months in a year, $t = 8/12$

$$A = P(1 + r\,t) = \$500\,(1 + .071 \cdot \frac{8}{12}) = \$523.67$$

$$\mathbf{I} = P\,r\,t = \$500(.071 \cdot \frac{8}{12}) = \$23.67 \text{ or } I = A - P = \$523.67 - \$500.00 = \$23.67$$

Example:

The owner of Lincoln Lovely Locks borrowed $12,000 in a 3% simple annual interest loan, and paid it back after 100 days. How much would the owner pay to clear the loan and what is the interest paid?

P = \$12000, r = .03, for time t, as there are 365 days in a year, then t = 100/365

$$A = P(1 + r\,t) = \$12000\left(1 + .03 \cdot \frac{100}{365}\right) = \$12098.63$$

$$I = P\,r\,t = \$12000\left(.03 \cdot \frac{100}{365}\right) = \$98.63 \text{ or } I = A - P = \$12098.63 - \$12000 = \$98.63$$

Practice

find the end balance and interest for these simple interest scenarios

a) Wilson borrowed \$32000 at a 5% annual interest for 6.5 years
End Balance owed ____________ Interest paid ________

b) The owner of Norbit Hair Care invested \$7000 in a 6.6% simple annual interest account and he cashed it in after 11 months.
End Balance earned ____________ Interest earned ________

c) Jennie borrowed \$5000 at 9.2% simple annual interest and he paid it back after 400 days.
End Balance owed ____________ Interest paid ________

Answers:

a) Wilson End balance = \$32000 (1 + .05 × 6.5) = \$42500, Interest= \$32000 (.05) (6.5) = \$10400,

b) Norbit end balance = \$7000 (1 + .066 × 11/12) = \$7423.50, Interest = \$7000 (.066) (11/12) = \$423.50,

c) Jennie end balance = \$5000 (1 + .092 × 400/365) = \$5504.11, Interest = \$5000 (.092) (400/365) = \$504.11

Many simple interest loans are paid back (or earned) at regular multiple times during the year, so this enters into the calculation of interest.

Here is the breakdown of times paid back during the year that are used

Time period	Number
Daily	365
Weekly	52
Semimonthly	24
Biweekly	26
Monthly	12
Quarterly	4
Semiannual	2
Annual	1

APR – ANNUAL PERCENTAGE RATE

Interest rates are usually given as an **annual percentage rate (APR)** – the total interest that will be paid in the year. If the interest is paid in smaller time increments, the APR will be divided up.

For example, a 6% APR paid monthly: as there are 12 months in a year, then 6%/12 = .5%, so this would be divided into twelve 0.5% charges or payments during the year.

A 4% annual rate paid quarterly: as this is 4 times in the year, and 4%/4 = 1%, this would be divided into four 1% charges or payments during the year.

A 13.5 % annual rate paid semiannually (twice a year): as 13.5%/2 = 6.75%, would be divided into two 6.75% charges or payments during the year.

Example

Treasury Notes (T-notes) are bonds issued by the federal government to cover its expenses. Suppose you obtain a $1,000 T-note with a 6% annual rate, paid semi-annually, with a maturity in 4 years. How much will the end balance be and how much interest will you earn?

P = $1000 as the 6% annual interest is paid semiannual, so 6%/2 = 3% so that means this will be divided into two 3% payments so r = .03

4 years means 4 × 2 = 8 semiannual payments. So t = 8

And A = P (1 + r t) = $1000 (1 + .03 · 8) = $1000 (1.24) = $1240

And the Interest is **I** = P r t = $1000 (.03) (8) = $240 (or I = A – P= $1240 – $1000 = $240)

Note: as it is simple interest based on an annual interest rate, you can also treat this as an annual simple interest problem, where the principal, P, is $1000, the annual interest rate, r, is 6% or .06, and the time, t, is 4 years.

So A = P (1 + rt) = 1000 (1 + .06 · 4) = 1000 (1 . 24) = $1240. You will get the same result.

Example:

Arnold borrowed $60000 to fund his Beauty Shop startup from a lender at an annual 8% simple interest that is charged quarterly. Arnold pays the loan after 5 years.

Let P = $60000 as the annual interest is 8% is charged quarterly, then 8% / 4 = 2% so there is four 2% charged during the year. So r = .02

5 year means 5× 4 = 20 quarterly charges. So t = 20.

And A = P (1 + r t) = $60000 (1 + .02 · 20) = $60000 (1.4) = $84000

And the Interest paid is **I** = P r t = $60000 (.02) (20) = $24000 (or I = A – P= $84000 – $60000 = $24000.)

And, you could also treat it in an annual interest rate way, let P = $60000, r = .08 and t = 5 and A = P (1 + rt) = $60000 (1 + .08 · 5) = $60000 (1 . 4) = $84000

Practice

Find the end balance and interest for these simple interest scenarios

a) Quinton Quality Beauty Care obtained $50000 Bonds at a 4.68 % annual interest paid monthly with a maturity of 6 years.

End Balance owed ________ Interest paid ________

b) Dolly invested $4000 in a 7.2% simple annual interest account that is paid semimonthly and he cashed it after 2 years.

End Balance earned ________ Interest earned ________

c) Ken Doll borrowed $34000 at 8.26% simple annual interest charged semiannually that he pays it back after 5 years.

End Balance owed ____________ Interest paid ____________

Answers:

a) Quinton Quality : P = $50000 monthly means 12 times a year, so 4.68%/12 = .38% so r = .0039, and 6 years means 12 x 6 = 72 charges, so t =72 end balance = $50000 (1 + .0039 · 72) = $64040, interest = $50000 (.0038) (72) = $114040; Or let P = $50000, r = .0468, t = 6 and end balance = 55000 (1 + .0468 · 6) = $64040;

b) Dolly : let P = $4000 Semimonthy means 24 times a year, so 7.2%/24 = .3 % so r = .003 and 3 years means 24 × 2 = 48 charges so t = 48, end balance = $4000 (1 + .003 × 48) = $4576, interest = $4000 (.003) (48) = $576; Or let P = $4000, r = .072, t = 2, and end balance = 4000 (1 + .072 · 2) = $4576

c) Ken Doll: P = $34000, Semiannual means 2 times a year so 8.26%/2 = 4.13% so r = .0413 and 5 years means 2 × 5 = 10 charges so t = 10 End balance = $34000 (1 + .0413 × 10) = $48042, interest = $34000 (.0413) (10) = $14042; Or let P = $34000, r = .0826 and t = 5, then end balance = 34000 (1 + .0826 · 5) = $48042

6-2 Assignments

Show all work!

1. Find the ending balance and the simple interest for the following principal, annual % interest and time.

	Principal	% rate	time	End Balance	Interest
a)	$35000	3.4	2 years	________	________
b)	$15000	5.6	4.5 years	________	________
c)	$4000	2.7	5 years 9 months	________	________
d)	$13500	8.88	400 days	________	________
e)	$23555	7	225 days	________	________

2. Needa Doe borrowed $250 from her friend Banksie in a SIMPLE interest loan at 10.2% annual interest. If Needa pays the loan back after 500 days, how much does she pay Banksie?

____

3. Julie borrowed $5000 from a lender to use for her hair salon remodeling. It was a simple interest loan at 6.7% annual interest. She paid it back after 2 years. What was Julie's payoff for her loan?

____

4. LuLu borrowed $800 from a colleague at 22.1% simple annual interest and paid it back after 10 months. How much does LuLu payback to the colleague ___________ and how much interest does LuLu have to pay ___________?

5. The Hair Shoppe borrowed $10000 in a 4.5% simple annual interest loan, and paid it back after 150 days. How much would The Hair Shoppe pay to clear the loan ___________ and what is the interest paid ___________?

6. Find the end balance and the simple interest for the following principal amounts, annual interest rate, time period charged, and time.

	Principal	% interest	period charged	time	end balance	interest
a)	$12400	5	Quarterly	4 years	________	________
b)	$35000	11.5	Monthly	3 years	________	________
c)	$675000	9.99	semiannually	8 years	________	________
d)	$567	4.5	daily	520 days	________	________
e)	$70050	3.3	annual	4 month	________	________

7. Timmie borrowed $4000 from a lender to fund his Boutique Shop inventory at an annual 5% simple interest that is charged weekly. Timmie pays the loan after 2 years. What does Timmie pay at the end of the loan to pay it off? ________________

8. Betty Beauty Care purchased $60000 in a bond at a simple 3.68% annual interest charged monthly with a maturity of 4 years. How much does Betty Beauty have in the investment after maturing in 4 years. ________________

9. Warren invested $4990 in a 7.2% simple annual interest account that is paid semimonthly and he cashed it after 6 years. What does Warren have after the 6 years? ________________

10. A principal of $5500 that is borrowed at 8.26% simple annual interest charged semiannually for 3 ½ years, what is the total amount of the <u>interest</u> that is charged? ________________

11. Petunia borrows $2200 in a simple 10.6% annual interest charged daily for 700 days. What does Petunia pay back after the 700 days? ________________

12. The owner of Worthington Beauty Supplies invested $3700 in a simple interest account at 7.2% annual interest that is paid quarterly. The time of the investment for maturity is 3 years. What does the owner collect after 3 years? ________________

6-3 COMPOUND INTEREST, SINGLE DEPOSIT TO END BALANCE

Let's again look at the T-Bill problems from earlier …

Treasury Notes (T-notes) are bonds issued by the federal government to cover its expenses. Suppose you obtain a \$1000 T-note with a 6% annual rate, paid semi-annually, with a maturity in 4 years. How much will the end balance be and how much interest will you earn?

We figured this with the 6% annual interest paid semiannually as two 3% payments every 6 months, and in 4 years, that means 8 total payments. So $P = \$1000$, $r = .03$, and $t = 8$.

And $A = P(1 + r\,t) = \$1000\,(1 + .03 \cdot 8) = \$1000\,(1.24) = \$1240$

We can also figure this as 6% annual interest with 4 yearly time period so $r = .06$, and $t = 4$ and

$A = P(1 + r\,t) = \$1000\,(1 + .06 \cdot 4) = \$1000\,(1.24) = \$1240$

This is because this is simple interest over time, and, with simple interest, we were assuming that we pocketed the interest when we received it. So interest is pocketed at 3% every six months or at 6% every year. And the Principal remains the same \$1000.

However, what is more common in loans or investments at lending facilities, banks and investment companies is any interest we accumulate or earn is automatically added to our balance, and we accumulate or earn interest on that interest in future years. This reinvestment of interest is called COMPOUNDING. And it results in an increase in either loan balance or investment balance dollars over simple interest over time!

KEY TAKAWAYS

- Simple interest is calculated using only the principal balance of the loan or investment
- With compound interest, the interest per period is based on the principal plus any interest already gained or charges. So interest compounds over time.

COMPOUND INTEREST ANNUALLY

Suppose we invest \$40000 in an account that earns 6% interest annually, but is compounded every year. So P = \$40000 and r = .06.

After one year, t = 1, and we have an ending balance $A = P(1 + rt) = 40000(1 + .06 \cdot 1) = 40000(1.06) = \$42,400$, and interest of \$2400 earned. This would be the same as if we apply simple interest over one year. Now, with compounding interest, the second year's 6% interest will be applied to the \$42,400 first year balance (and not \$40,000, as with simple interest) then added to the \$42,400 to reach the new ending balance after two years.

So for the second year, P = \$42,400, r = .06 and t = 1, and

$A = P(1 + rt) = \$42400(1 + .06 \cdot 1) = \$42400(1.06) = \$44944$, and total interest for the two years of \$44944 – \$40000 = \$4944.

(Note: with simple interest of 6% over two year, the ending balance would be $40000(1 + .06 \bullet 2) = \$44800$, which is \$144 less than compounding yearly . .)

That means the third year is $A = \$44944(1 + .06 \cdot 1) = \$47640.64$

And the fourth year is $A = \$47640.64(1.06 \cdot 1) = \$50499.08$, and so on ....

Let's look at want is going with compound interest done annually. As t is 1,

1st year end balance = $P(1 + r)$

2nd year end balance = (1st year end balance) $(1 + r) = P(1 + r)(1 + r) = P(1 + r)^2$

3rd year end balance =(2nd year end balance)$(1 + r) = P(1 + r)^2(1+r) = P(1 + r)^3$

So, Following this pattern, then 4th year end balance = $P(1 + r)^4$

then 5th year end balance = $P(1 + r)^5$ and so on

We can figure a formula for annual interest that is compounded annually as

A = end balancc P = Principal, r = annual interest, m = the number of years of compounding

$$\mathbf{A = P(1 + r)^{m}}$$

Example:

Wellington Wellness Center invested \$50000 in a 7.7% annual interest account compounded annually. What is in the account after 10 years? And How much interest does Wellington earn?

P = \$50000, r = .077 m = 10 so $A = P(1 + r)^m = \$50000(1 + .077)^{10} = \$104984.95$

And interest earned, $I = A - P = \$104984.95 - \$50000 = \$54984.95$

Example:

SallyMae took out a loan compounded annually with a principal of \$4000 at an annual interest rate of 2.5%. She pays it back after 3.5 years. What does SallyMae's pay back after 3.5 years? And how much interest does she pay in all?

P = \$4000 r = .025 t = 3.5 so $A = P(1 + r)^m = \$4000(1 + .025)^{3.5} = \$4361.07$ which equals her payout.

And interest paid is \$4361.07 – \$4000.00 = \$361.07

Practice

Find the Ending balance and interest for the following money invested or borrowed that has the interest compounded annually

	Ending balance	Interest earned
a) \$24500 at 8% interest for 6 years	________	________
b) \$135000 at 4.5% for 25 years	________	________
c) \$11110 at 11% for 12.5 years	________	________

Answer:

a) Ending Balance = \$24500.00 $(1 + .08)^6$ = \$38878.42, Interest = \$38878.42 – \$24500.00 = \$14378.42

b) Ending balance = \$135000.00 $(1 + .045)^{25}$ = \$405733.65, Interest = \$405733.65 – \$135000 = \$270733.65

c) ending balance = \$11110.00 $(1 + .11)^{12.5}$ = \$40949.75, Interest = \$40949.75 – \$11110.00 = \$29839.75

COMPOUND INTEREST OTHER TIME PERIODS

What about investments or loans that have an annual interest rate, but are compounded more frequently that once a year. You find investments or loans that are set up to compound interest twice a year, quarterly, monthly, weekly, or daily (and other time periods.)

Here is a table showing are the different periods in year for compounding schedules.

Time compound schedule	Compounding number in a year
Daily	365
Weekly	52
Semimonthly	24
Biweekly	26
Monthly	12
Quarterly	4
Semiannual	2
Annual	1

How will this affect the overall end balance? Let's apply the same thinking as we did for annual compounding.

Example:

Louie invested \$1000 in an account at 9% annual interest that is compounded 12 times a year, or monthly. After 20 years, how much would Louie have in his account, and how much interest would he earned?

As there are 12 months in a year, that means the 9% annual interest would be broken up as 9% divided by 12 = $\frac{9}{12}\%$ = .75%, so .75% is applied every month period.

So for the first month end balance, P = \$1000, r = .09, so $\frac{r}{12}\% = \frac{9}{12}$ = .75% = .0075

and $A = P\left(1 + \frac{r}{12}\right)$ = \$1000 (1.0075) = \$1007.50.

The Second month end balance A = \$1007.50 (1.0075) = \$1015.06

The third month end balance A = \$1015.06 (1.0075) = 1022.67 and so on.

This works out as

1st month end balance = $P\,(1 + \frac{r}{12})$

2nd month end balance = (1st month end balance) $(1 + \frac{r}{12}) = P\,(1 + \frac{r}{12})\,(1 + \frac{r}{12}) = P\,(1 + \frac{r}{12})^2$

3rd year month balance = (2nd month end balance) $(1 + \frac{r}{12}) = P\,(1 + \frac{r}{12})^2\,(1+\frac{r}{12}) = P\,(1 + \frac{r}{12})^3$

So, Following this pattern, then 4th month end balance = $P\,(1 + \frac{r}{12})^4$

5th month end balance = $P\,(1 + \frac{r}{12})^5$ and so on

We can figure a formula for annual interest that is compounded **MONTHLY** as

A = end balance P = Principal, r = annual interest, m = the number of **months** of compounding, then

$$A = P\,(1 + \frac{r}{12})^m$$

So for Louie, P = \$1000, r = .09, 20 years is 12 × 20 = 240 months of compounding, and

Ending balance after 20 years A = \$1000 $(1 + \frac{.09}{12})^{240}$ = \$6009.15

And the interest earned = \$6009.15 –\$1000.00 = \$5009.15

What if the investment was compounded 4 times a year, or QUARTERLY. If the same problem had compounding QUARTERLY instead of monthly, then the annual r percentage interest would be divided into 4 equal percentages applied four times in the year, so we would have $\frac{r}{4}$% applied every 3 months. So out formula for annual interest that is **QUARTERLY** compounding with

A = end balance P = Principal, r = annual interest, m = the number of **quarters** of compounding and

$$A = P\,(1 + \frac{r}{4})^m$$

So if our Louie problem was compounded quarterly, then

P = \$1000, r = .09, m = 20 years × 4 = 80 quarters of compounding and

Ending balance after 20 years A = \$1000 $(1 + \frac{.09}{4})^{80}$= \$5930.15

Can you see that the number that the annual rate r is divided by in the formula is the number of compounding time periods during the year.

Given the different possible compounding ways in a year, here are the rate r/(number of compound periods in a year)

Time compound schedule	r/(Number of compound periods in a year)
Daily	r/365
Weekly	r/52
Semimonthly	r/24
Biweekly	r/26
Monthly	r/12
Quarterly	r/4
Semiannual	r/2
Annual	r/ 1 = r

We can generalize our formula as follows

Let A = end balance, P = Principal, r = annual interest, k is number of compounding periods in a year, N = the number of years of compounding, so $N \cdot k$ = the number of total compounding period, then

$$\mathbf{A = P\,(1 + \frac{r}{k})^{Nk}}$$

Example:

Sheila has a savings account is compounded daily, at 4.5% annual interest. If $4500 is deposited and left for 8 years, how much is in the account at the end of the 8 years? And how much interest does she earn on the savings?

P = $4500, r = .045, k = 365, N = 8, so

$$A = P\,(1 + \frac{r}{k})^{Nk} = \$4500\,(1 + \frac{.045}{365})^{8 \cdot 365} = \$4500\,(1 + \frac{.045}{365})^{2920} = \$6449.84 \text{ end balance}$$

Interest earned is $6449.84 – $4500.00 = $1949.84

Example:

Anastasia borrowed $49500 from a bank for repairs on her beauty shop. The loan was at 5.58% annual interest compounded semiannually. She pays the loan back after 6.5 years. What Anastasia's pay out on the loan? And how much interest does she pay?

P = $49500, r = .0558 k = 2 N = 6.5

$$A = P\,(1 + \frac{r}{k})^{Nk} = \$49500\,(1 + \frac{.0558}{2})^{6.5 \cdot 2} = \$49500\,(1 + \frac{.0558}{2})^{13} = \$70789.11 \text{ end balance}$$

And interest paid is $70789.11 – $49500.00 = $21289.11

Practice

a) Willie put \$23000 into a savings account earning 6.66% annual interest compounded quarterly for 8 years. What is in the account after the 8 year? ____________________
How much interest did Willie earn? ____________________

b) Zelda invested \$2400 into an account earning 5% annual interest compounded weekly for 5.5 years. What is in the account after 5.5 years? ____________________
How much interest was attained? ____________________

c) To save money for her future Hair Salon down payment, Doris invested \$2500 in an account compounded biweekly at 8.9% annual interest for 13 years. What was in the account after 13 years? ____________________

d) Kent took out a loan of \$560 at 9% annual interest compounded daily for 221 days. How much did Kent payout after the 221 days? ____________________. How much interest was paid out? ____________________

Answers:

a) $\$23000\,(1 + .0666/4\,)^{8 \cdot 4} = \$39013.48$; $\$38829.71 - \$23000.00 = \$16013.48$

b) $\$2400\,(1 + .05/52\,)^{5.5 \cdot 52} = \$3159.26$; $\$3159.26 - \$2400.00 = \$759.26$

c) $\$2500\,(1 + .089/26\,)^{26 \cdot 13} = \$7935.25$

d) d) $\$560\,(1 + .09/365\,)^{\frac{221}{365} \cdot 365} = \$560\,(1 + .09/365\,)^{221} = \$591.36$; $\$591.38 - \$560.00 = \$31.36$

6-3 Assignments

Show all work !!

1. Find the ending balance and interest for the following principal, annual interest, compounding period, and time

	Principal	% interest	Compound period	time	End Balance	Interest
a)	$100400	5.5	Quarterly	3 years	________	________
b)	$37000	10.5	Monthly	2.5 years	________	________
c)	$675000	2.99	semiannually	14 years	________	________
d)	$16000	3.5	daily	250 days	________	________
e)	$6000	8.3	annual	5 months	________	________
f)	$5340	4.35	weekly	17 years	________	________

2. Lottie Monet had an investment of $450000 in a 4.5% annual interest account, compounded monthly. Lottie leaves this money to grow 27 years. What is in her balance after 27 year?

____

3. Dorothy invests $23000 in an account compounded quarterly for 30 years for retirement. If the annual interest percentage is 6.7%, how much does Dorothy collect after the 30 years?

____

4. Alice puts $44800 into a savings account earning 5.66% annual interest compounded monthly for 8 years. What is in the account after the 8 year? ____________________

5. Zelda invested $11000 into an account earning 9.7% annual interest compounded weekly for 4 years. What is in the account after the 4 years? ____________________

6. To save money for her future Hair Salon down payment, MaggieMae invested $3100 in an account compounded biweekly at 9.9% annual interest for 10 years. What was in the account after 10 years? ____________________

7. Clark Kent invested $960 in an account at 8% annual interest compounded daily for 641 days. How much was in the account after the 641 days? ____________________

8. Andy Taylor invested $30000 in the Mayberry Sheriff Credit Union at an annual interest of 5.99% compounded monthly. He leaves the account going for 25 years. How much is in his credit union account after the 25 years? ____________________. How much interest did Andy earn? ____________________

9. Aunt Bea ran a hair care shop in Mt. Pilar. She wants to know which investment would bring in the most money after 20 years.

A $2000 deposit in a 7% annual interest account compounded annually or

A $1450 deposit in an 8.5% annual interest account compounded monthly

The ____________ account brings in ____________ more at the end of 20 years.

10. Figure: An investment of $50000 at 6.99% annual interest for 35 years will create a ending balance for the following compound periods

a) Annual ____________________

b) Semiannual ____________________

c) Quarterly ____________________

d) Monthly ____________________

e) Semi Monthly ____________________

f) Biweekly ____________________

g) Weekly ____________________

h) Daily ____________________

11. Bruce Wayne puts up $176,000 in an investment at 4.125% annual interest at the Gotham City Bank for 10 years. How much more money is earned in this investment if Bruce can get an account that is compounded monthly rather than compounded annually?

End Balance compounded monthly ______________

End Balance compounded annually ____________

Amount gained by Monthly over Annual compounding ___________________

6-4 COMPOUND INTEREST, END BALANCE TO SINGLE DEPOSIT

In business or personal planning, we may have a goal of how much money is needed after so many months of investing. So for a future balance needed on a compound investment, if we know the annual interest rate, the compound period, the number of years of investing, then we would want to know how much in principal needs to be deposited at the beginning to attain the desire end balance.

We will simply manipulate the compounding formula in order to solve the formula for the Principal P, so

Given: A = end balance, P = Principal, r = annual interest, k is number of compounding periods in a year, N = the number of years of compounding, so N · k = the number of total compounding periods then

$$P\left(1+\frac{r}{k}\right)^{Nk} = A$$

To solve it for P

Dividing both sides of our formula equation by $\left(1+\frac{r}{k}\right)^{Nk}$ we get

$$P\left(1+\frac{r}{k}\right)^{Nk} \div \left(1+\frac{r}{k}\right)^{Nk} = A \div \left(1+\frac{r}{k}\right)^{Nk}$$

or

$$\mathbf{P} = \frac{\mathbf{A}}{\left(1+\frac{r}{k}\right)^{Nk}}$$

We can use this now to figure the Principal needed.

Example:

Terri want to have $35000 after 8 years to put toward her beauty business. She finds an annual compoundinvestment at 7.4% annual interest. What does she need to deposit in the account to meet her goal after 10 years.

Let A = $35000, r = .074, k= 1, N = 10 and

$$P = \frac{A}{\left(1+\frac{r}{k}\right)^{Nk}} = \frac{\$35000}{\left(1+\frac{.074}{1}\right)^{10\cdot 1}} = \$17140.57 \text{ needed to be deposited at the start.}$$

Example:

Ethan finds a monthly compound investment at 9.77% annual interest. Ethan wants to raise $150000 after 20 years. How much does Ethan need to put down as principal to meet this goal?

Let A = $150000, r = .0977, k = 12, N = 20 so

$$P = \frac{A}{\left(1+\frac{r}{k}\right)^{Nk}} = \frac{\$150000}{\left(1+\frac{.0977}{12}\right)^{20\cdot 12}} = \$21424.75 \text{ needed to be deposited at the start.}$$

Practice

a) Yolanda needed to save $30000 in 5 years. If she invested in a 11.5% annual interest compounded quarterly account, what does she need to put down to meet this goal? ___________

b) If one invests in a 3.6% annual interest account that is compounded daily for 700 days, and there is $1250 in the account, what was the initial principal? ___________

c) Holly's Hair invested some money in a 12.5% annual interest account compounded weekly. After 4 years there is $41500 in the account. What was originally put down to achieve this ending balance after 4 years? ___________

d) An investment compounded biweekly for 12 years at 13.5% annual interest. If there was $250000 in the account after 12 years, what was the initial principal deposit? ___________

Answers:

a) $\$30000/(1 + .115/4)^{5 \cdot 4} = \$17018.62$

b) $\$1250/(1 + .036/365)^{\frac{700}{365} \cdot 365} = \$1250/(1 + .036/365)^{700} = \$1166.61$

c) $\$41500/(1 + .125/52)^{4 \cdot 52} = \$25186.13$

d) $2500000/(1 + .135/26)^{12 \cdot 26} = \$49682.47$

6-4 Assignments

Show all work!!

1. Find the single principal deposit needed for the following annual interest, compounding period, time, and end balance needed

	% interest	Compound period	time	End Balance	Principal
a)	5.5	quarterly	7 years	$200400	________
b)	8.65	monthly	9.5 years	$310000	________
c)	3.3	weekly	18 years	$195000	________
d)	11.5	annually	17 years	$458000	________
e)	9.98	daily	290 days	$1340	________
f)	16.5	semiannual	5 years	$36000	________

2. Gilly Bates wants to have an account worth \$2300000 after 25 years. If Gilly has a monthly compounding account at 7.77% annual interest, what single principal deposit does he need to make to accomplish this?

____

3. DeeDee invests in a 6.7% annual interest investment for 12 years that is compounded quarterly. Her goal is \$280000 total, so what does DeeDee need to deposit at the start of the account?

____

4. Wilson puts some principal into a savings account earning 6.66% annual interest compounded weekly for 8 years and had \$17800 saved up. What was Wilson's single deposit amount?

____

5. Zorro plans to have \$250000 after 22 years in a single deposit account that is compounded bimonthly at 12.5% annual interest means how much of a single principal deposit is needed at the beginning of the investment? ____________________

6. For her future Health Spa down payment, MindySue needs \$10000. She finds a compounded biweekly investment at 9.9% annual interest. If she plans on investing for 10 years, what principal is needed? ____________________

7. Deputy Dawg invested some deposit money in a daily compounding account at 17% annual interest for 800 days. If there was \$3450 in the account after the 800 days, how much did Deputy Dawg initially deposit? ____________________

8. What investment plan with an end balance of \$2000 after 5 years requires the largest single deposit and by how much more

Plan A: 8% annual interest compounded monthly or

Plan B: 7.8% annual interest compounded weekly

____

ANNUITIES

When saving for business or personal reasons, one often makes a regular deposit to the account over time. The is called an annuity.

An **annuity** is an investment in which the purchaser makes a sequence of periodic, equal payments. To find the amount of an annuity, we need to find the sum of all the payments and the interest earned. This is a very common way to save, such as having a portion of your monthly paycheck put into a retirement account. Each time a payment is paid, interest is applied to the growing balance in the account that is compounded over period. Over time, the account grows at a rapid rate.

An annuity differs from a one-time compound interest account. Here are the distinctions.

Annuities assume that you put money in the account on a regular schedule (every month, year, quarter, etc.) and let it sit there earning interest.

Compound interest assumes that you put money in the account once and let it sit there earning interest. So . . . . .

Compound interest: One deposit

Annuity: Many deposits.

6-5 ANNUITIES, REGULAR DEPOSITS TO END BALANCE

Here is an example to illustrate how an annuity works.

$1000 is deposited at the beginning of each year at 5% annual interest, compounded annually. What is in the account after the 20th year deposit?

First year:

$1000 is deposited at the beginning of a year, and after the first year there is $1000 (1 + .05) = $1050 in the ending balance after the first year.

Second year:

We start the second year with $1050 plus another $1000 deposited, or $1050 + $1000 = $2050. Compound interest for this amount is $2050 (1 + .05) = $2152.50, the ending balance after the second year.

Third year:

We start the third year with $2152.50 plus another $1000 deposited, or $2152.50 + $1000 = $3152.50. Compound interest for this amount is $3152.50 (1 + .05) = $3310.13, the ending balance after the third year, and so on . . . .

This would take some time to calculate this out for 20 years. However, to calculate an Annuity end balance with a regular deposit paid at the beginning of an annualtime period, we can use the following formula:

Given: d = the deposit amount each period, r = the annual rate, N the number of years of investment, and A_N = ending balance after the Nth year deposit (that is at the beginning of the Nth year)

$$A_N = \frac{d\,((1+r)^N - 1)}{r}$$

Let look at our example, Let d = \$1000, r = .05, N = 1

So $A_{beginning\ 1st\ year} = \dfrac{\$1000\,((1+.05)^1 - 1)}{.05} = \$1000,$

Now, let N = 2 So $A_{beginning\ 2nd\ year} = \dfrac{\$1000\,((1+.05)^2 - 1)}{.05} = \$2050$

Let N = 3 so $A_{beginning\ 3rd\ year} = \dfrac{\$1000\,((1+.05)^3 - 1)}{.05} = \$3152.50$

You should see these match our results we got at the beginning of this section.

To figure the end balance after the 20th year deposit, Let N = 20

and $A_{beginning\ 20th\ year} = \dfrac{\$1000\,((1+.05)^{20} - 1)}{.05} = \$33065.95.$

So there would be \$33065.95 in the annuity account after the 20th deposit.

Example:

An annuity account is set up with \$500 deposited at the beginning of each year at an annual interest rate of 7.4% compounded annually. What is in the ending balance after the 15th year deposit.

d = \$500 r = .074 N = 15

$$A_{beginning\ 15th\ year} = \frac{500\,((1+.074)^{15} - 1)}{.074} = \$12958.50$$

What about deposits made at <u>different</u> regular periods over the year? To generalize this annuity formula to find compounding over <u>different time period,</u> our formula is

Given: d = the deposit amount each period, r = the annual rate, k is the number times deposited in a year as well as the number of compounding period in a year, N the number of years of investment, and $A_{N \cdot k}$ = ending balance after the $(N \cdot k)$ deposits

$$A_{N \cdot k} = \frac{d\left((1+\frac{r}{k})^{N \cdot k} - 1\right)}{\left(\frac{r}{k}\right)}$$

Example:

An annuity calls for a monthly deposit of \$450 at the beginning of each month at an annual 4.5% interest compounded monthly, done for 10 years, What is the balance after the last deposit.

d = \$450, r = .045, k = 12, N = 10

$$A_{N \cdot k} = \frac{d\left(\left(1+\frac{r}{k}\right)^{N \cdot k}-1\right)}{\left(\frac{r}{k}\right)} = \frac{\$450\left(\left(1+\frac{.045}{12}\right)^{10 \cdot 12}-1\right)}{\left(\frac{.045}{12}\right)} = \$\,68039.13$$

Example:

An annuity calls for a deposit of \$10000 twice a year at an annual 5.15% interest compounded semiannually for 20 years. What is the balance of this account after the last deposit?

d = \$10000, r = .051, k = 2, N = 20

$$A_{N \cdot k} = \frac{d\left(\left(1+\frac{r}{k}\right)^{N \cdot k}-1\right)}{\left(\frac{r}{k}\right)} = \frac{\$10000\left(\left(1+\frac{.0515}{2}\right)^{20 \cdot 2}-1\right)}{\left(\frac{.0515}{2}\right)} = \$685352.56$$

Example:

Jericho Joe put \$25.00 daily into an account compounded daily at an annual interest of 2.8%. After 2 years, how much did Jericho Joe have in the account?

d = \$25, r = .028, k = 365, N = 2

$$A_{N \cdot k} = \frac{d\left(\left(1+\frac{r}{k}\right)^{N \cdot k}-1\right)}{\left(\frac{r}{k}\right)} = \frac{\$25\left(\left(1+\frac{.028}{365}\right)^{2 \cdot 365}-1\right)}{\left(\frac{.028}{365}\right)} = \$\,18769.93$$

PRACTICE PROBLEMS

a) Holly takes out an annuity where \$2500 is deposited at the beginning of each year for 30 years. If the annual interest rate is 8.5% compounded annually, what is the balance after the 30th deposit?

____

b) An annuity is set up where \$6890 is deposited at the beginning of each quarter for 10 years. If the annual interest rate is 4.52% compounded quarterly, what is the balance after the last deposit?

____

c) An annuity is set up where \$750 is deposited at the beginning of each week for 20 years . If the annual interest rate is 3.99% compounded weekly, what is the balance after the last deposit?

____

d) In an annuity, \$30 is deposited at the beginning of each day for 12 years. If the annual interest rate of 9.9% is compounded daily, what is in the balance after the last deposit?

____

Answers:

a) $$\text{end balance} = \frac{\$2500((1+\frac{.085}{1})^{30\cdot 1}-1)}{\left(\frac{.085}{1}\right)} = \$310536.81$$

b) $$\text{end balance} = \frac{\$6890((1+\frac{.0452}{4})^{10\cdot 4}-1)}{\left(\frac{.0452}{4}\right)} = \$346008.36$$

c) $$\text{end balance} = \frac{\$750((1+\frac{.0399}{52})^{50\cdot 52}-1)}{\left(\frac{.0399}{52}\right)} = \$\ 1192886.59$$

d) $$\text{end balance} = \frac{\$30((1+\frac{.099}{365})^{12\cdot 365}-1)}{\left(\frac{.099}{365}\right)} = \$252180.18$$

6-5 Assignments

Show all work!!

1. For these annuities, figure the end balance

Deposit amount	deposit mode	% interest	compound period	time	End Balance
a) $500	annually	5%	annually	15 years	________
b) $35	weekly	9.8%	weekly	8 year	________
c) $1000	monthly	7.65%	monthly	28 years	________
d) $25	daily	5.5%	daily	40 years	________
e) $2875	Quarterly	2.55%	quarterly	10.5 years	________

2. Sam takes out an annuity where $300 is depositedmonthly for 30 years. If the annual interest rate is 9.5% compounded monthly, what is Sam's balance after the last deposit? ____________

3. An annuity is set up where $1210 isdeposited at the beginning of each quarter for 8.5 years. If the annual interest rate is 9.52% compounded quarterly, what is the balance after the last deposit? ____________

4. Sandy wants to save for retirement, so she sets up an annuity is set up where $95 is deposited at the beginning of each week for 30 years. If the annual interest rate is 5.99% compounded weekly, what is the balance of her account after the last deposit? ____________

5. In an annuity, $15 is deposited at the beginning of each day for 600 days. If the annual interest rate of 10.9% is compounded daily, what is in the balance after the last deposit? ____________

6. An annuity account is set up with $500 deposited at the beginning of each year at an annual interest rate of 7.4% compounded annually. What is in the ending balance after 18 years? ____________

7. An annuity calls for a semiannual deposits of $10000 at an annual 5.15% interest compounded semiannually for 40 years, What is the balance after the 40 years? ____________

8. Barney Fife set up an annuity at his Mayberry Credit Union where he deposits $1430 monthly for 30 years. If the annual interest rate is 7.99% compounded monthly what is Barney's balance after 30 years? ____________

9. Horatio deposited $450 Quarterly into an annuity account earning 8.5% interest compounded quarterly. How much will Horatio have in the account if he did this for 15 years? ____________

10. An annuity is set up where $210 is deposited biweekly for 28 years. If the annual interest rate is 9.52% compounded biweekly, what is the balance after the 28 years? ____________

11. Which savings program provides the most end balance and by how much?
    - An annuity of $450 deposited monthly at 5.6% annual interest compounded monthly for 25 years or
    - a single principal of $72500 invested at 5.6% annual interest compounded monthly for 25 years

______________________________ ____________

6-6 ANNUITIES: FINDING THE REGULAR DEPOSIT ON AN ANNUITY FOR A FUTURE END BALANCE

In planning for the future for a saving balance for a business down payment, purchasing a large item personal thing, or for retirement, it is often useful to figure the regular deposit one needs to make to achieve a desirable end balance amount. To figure this deposit amount, we do a variation on the annuity formula.

Remember,theannuityformulaisgivenas:

Given: d = the deposit amount each period, r = the annual rate, k is the number of compounding period in a year, N the number of years of investment, and $A_{N \bullet k}$ = ending balance after the $(N \cdot k)$ deposits then

$$A_{N \cdot k} = \frac{d((1+\frac{r}{k})^{N \cdot k} - 1)}{\left(\frac{r}{k}\right)}$$

We will manipulate the formula in order to "solve" this formula for the deposit, d.

If we multiply both sides by $\frac{r}{k}$ then divide both sides by $(1+\frac{r}{k})^{N \cdot k} - 1$

We get

$$A_{N \cdot k} \cdot \frac{\frac{r}{k}}{(1+\frac{r}{k})^{N \cdot k} - 1} = \frac{d((1+\frac{r}{k})^{N \cdot k} - 1)}{\left(\frac{r}{k}\right)} \cdot \frac{\frac{r}{k}}{(1+\frac{r}{k})^{N \cdot k} - 1}$$

Or

$$\frac{A_{N \cdot k}\left(\frac{r}{k}\right)}{(1+\frac{r}{k})^{Nk} - 1} = \mathbf{d}$$

So, to find the deposit amount, d, we use

$$\mathbf{d} = \frac{A_{N \cdot k}\left(\frac{r}{k}\right)}{(1+\frac{r}{k})^{Nk} - 1}$$

Example:

Regina wants to save \$70000 in 9 years with plans to put it toward her business. She finds a 4.5% annual interest compounded monthly. What would Regina need to deposit each month to meet this need?

$A_{N \cdot k}$ = \$70000, r = .045, k= 12, and N = 9, so

$$d = \frac{A_{N \cdot k}\left(\frac{r}{k}\right)}{(1+\frac{r}{k})^{Nk} - 1} = \frac{70000\left(\frac{.045}{12}\right)}{(1+\frac{.045}{12})^{9 \cdot 12} - 1} = \$526.93$$

Regina needs to deposit \$526.93 each month.

Example:

Terry Blewealthy goal is to raise \$300000 in an annuity after 25 years. He wants a quarterly annuity payment and finds one at 5.99% annual interest compounded quarterly. How much should Terry deposit every quarter to meet this goal?

$A_{N \cdot k}$ = \$300000, r = .0599, k = 4, N= 25, so

$$d = \frac{A_{N \cdot k}\left(\frac{r}{k}\right)}{(1+\frac{r}{k})^{Nk} - 1} = \frac{\$300000\left(\frac{.0599}{4}\right)}{(1+\frac{.0599}{4})^{25 \cdot 4} - 1} = \$1313.16$$

So Terry needs a deposit of \$1313.16 every 3 months (quarter)

Example:

Hedda goal is to have \$1000000 in retirement after 35 years. She finds an annuity that has an 8.7% annual interest compounded semiannually. What does Hedda need to contribute every 6 months to achieve this goal?

$A_{N \cdot k}$ = \$1000000, r = .087 k = 2, N = 35

$$d = \frac{A_{N \cdot k}\left(\frac{r}{k}\right)}{(1+\frac{r}{k})^{Nk} - 1} = \frac{\$1000000\left(\frac{.087}{2}\right)}{(1+\frac{.087}{2})^{35 \cdot 2} - 1} = \$2326.17$$

So \$2326.17 needs to be deposited every 6 months (semiannually.)

Example:

Lisa needs a $12000 down payment for a house, and wants to raise it in 2 ½ years. She finds an annuity at 9.99% annual interest compounded weekly. What does she need to deposit every week to raise this amount?

$A_{N \cdot k} = \$12000$, r = .0999, k = 52, N = 2.5

$$d = \frac{A_{N \cdot k}\left(\frac{r}{k}\right)}{(1+\frac{r}{k})^{Nk} - 1} = \frac{\$12000\left(\frac{.0999}{52}\right)}{(1+\frac{.0999}{52})^{52 \cdot 2.5} - 1} = \$81.35$$

So $81.35 deposited every week.

PRACTICE PROBLEMS

a) Lonnie purchased an annuity at an annual interest rate of 3.4% compounded annually. To reach $79000 after 8 years, how much should Lonnie deposited every year? ____________________

b) Lottie Monie, an investor, wants to reach $400000 in a retirement account after 30 years. She has an annuity at 6.6% annual interest compounded monthly. What should Lottie deposit into her account every month to reach her retirement goal? ____________

c) An 10.3 % annual interest annuity that is compounded weekly resulted in $35000 after 10 years. What was the weekly deposit into the annuity? ______________

d) Biggie Spender had accumulated $21500 in an annuity compounded daily at 7.2% annual interest after 6 years. What was Biggie's daily deposit into the account? _________________

Answers:

a) $$d = \frac{A_{N \cdot k}\left(\frac{r}{k}\right)}{(1+\frac{r}{k})^{Nk} - 1} = \frac{\$79000\left(\frac{.034}{1}\right)}{(1+\frac{.034}{1})^{8 \cdot 1} - 1} = \$8758.74$$

b) $$d = \frac{A_{N \cdot k}\left(\frac{r}{k}\right)}{(1+\frac{r}{k})^{Nk} - 1} = \frac{\$400000\left(\frac{.066}{12}\right)}{(1+\frac{.066}{12})^{30 \cdot 12} - 1} = \$354.64$$

c) $$d = \frac{A_{N \cdot k}\left(\frac{r}{k}\right)}{(1+\frac{r}{k})^{Nk} - 1} = \frac{\$35000\left(\frac{.103.}{52}\right)}{(1+\frac{.103}{52})^{10 \cdot 52} - 1} = \$38.55$$

d) $$d = \frac{A_{N \cdot k}\left(\frac{r}{k}\right)}{(1+\frac{r}{k})^{Nk} - 1} = \frac{\$21500\left(\frac{.072}{365}\right)}{(1+\frac{.072}{365})^{6 \cdot 365} - 1} = \$7.85$$

6-6 Assignments

Show all work

1. For these ending balances, deposit mode, annual interest rate, compounding period, and time, find the regular deposit amount.

Ending balance,	deposited	% interest	compound period	time	DEPOSIT
a) $400000	monthly	6%	monthly	17 yr	________
b) $56000	quarterly	15%	quarterly	23 yr	________
c) $660000	semiannually	8.25%	semiannually	14.5 yrs	________
d) $1300000	annually	6.7%	annually	30 yr	________
e) $3450	daily	11.5%	daily	6 yr	________

2. Wussell Rilson wanted to have $230000 in a retirement account after 30 years. If he can earn 6.6% compounded monthly in an annuity account, how much does Wussel need to deposit at the end of each month to achieve this retirement goal? ____________

3. Mellisa wants to save $100000 in 12 years with plans to put it toward her retirement. She finds a 7.5% annual interest compounded semiannually. What would Regina need to deposit every 6 months to meet this need?

4. David's goal is to raise $300000 in an annuity after 20 years. He wants a monthly annuity payment and finds one at 10.95% annual interest compounded monthly. How much should David deposit every month to meet this goal? ____________

5. Mr. McScrooge's goal is to have $1500000 in retirement after 35 years. He finds an annuity that has a 5.7% annual interest compounded quarterly. What does Mr. McScrooge need to contribute every 3 months to achieve this goal? ____________

6. Barney Rubble needs a $10000 down payment for a house he wants to buy in Bedrock, and wants to raise it in 3 years. He finds an annuity at 12.99% annual interest compounded weekly. What does Barney need to deposit every week to raise this amount? ____________

7. Jeathro purchased an annuity at an annual interest rate of 5.4% compounded annually. To reach \$82000 after 9 years, how much should Jeathro deposit each year? _______________

8. Ina DeMonie, an investor, wants to reach \$500000 in a retirement account after 30 years. She has an annuity at 8.6% annual interest compounded monthly. What should Lottie deposit into her account every month to reach her retirement goal? _______________

9. A 10.3 % annual interest annuity compounded biweekly resulted in \$350000 after 40 years. What was the biweekly deposit into the annuity? _______________

10. Mr. Banks had accumulated \$21500 in an annuity compounded daily at 11.2% annual interest after 4.5 years. What was Mr. Banks's daily deposit into the account? _______________

11. Betty Boop needs \$25000 after 7 years. She deposited a regular amount twice a month at 14.55% annual interest compounded semimonthly. What was her regular deposit amount? _______________

CHAPTER 7

Preparing Payroll, Federal Taxes and Deductions

When running your business, you may have employees. Understanding how to figure taxes and deductions when doing payroll is important. Here are some key things to know:

7-1 GROSS INCOME

GROSS INCOME is all money due to an employee before any deduction. Gross income is usually considered "the salary" of an employee when asked how much does one make. For example, if an employee makes $15 per hour and works 40 hours, their gross pay would be 15 × 40 = 600 or $600 gross income.

Because of dollars deducted from the gross pay for tax, dues, benefits, and so on, the amount received by the employee is less than the gross pay. When money is deducted from the gross pay, the result is considered the employee's NET PAY. Net pay sometimes referred to as the "take-home pay" as it is what the employee gets to keep, or takes home with him or her, to use as they please.

For example: Here is a statement of earnings for an employee:

July 1, 2020–July 14, 2020	ACTIONJACKSON MOVIE COMPANY	
Employee: Seven Steegall	Position: Actor	Rate $37.50 / hour
Pay $	Deductions $	
Hours: 35	Federal Income Tax	140.15
Gross pay: $1312.50	Medicare Insurance	19.03
	Social Security Insurance	81.34
	Medical Insurance plan	21.80
	--	
	Total Deductions	262.32
NET PAY $1050.18		

The gross pay is figured by multiplying the hours times the hourly wage, 35 × $37.50 = $1312.50

Adding up all the deductions, we get $262.32. So the net pay is the difference between gross pay and deductions, or $1312.50 – $262.32 = $1050.18. So Seven "takes home" $1050.18.

PAYMENT PERIODS

Employees can be paid in any number of ways and times, but usually, people are paid their net pay either weekly, biweekly (every two weeks), semimonthly (twice a month), or monthly.

Biweekly and semimonthly are very close to the same, however, as there are 52 weeks in a year, biweekly would figure as 52/2 = 26 payments a year. As there are 12 months in a year, semimonthly would figure as 12 × 2 = 24 payments a year. So the same gross pay done biweekly and done semimonthly figures out differently.

For example:

if the gross pay is $1450 biweekly, it would figure as an annual salary of 26 × 1450 = $37,700. If gross pay of $1450 is paid semimonthly, it figures as an annual salary of 24 × 1450 = $34,800.

MINIMUM WAGE

Each state has set a certain minimum hourly wage for their workers. For example, the state of Washington has determined that workers are to be paid a minimum of \$13.50 per hour (January 1, 2020). Some Cities and counties jurisdictions have increased that rate.

Overtime is paid if an employee works in excess of 40 hours in a week. Employees are compensated at 1.5 times (or a 50% increase) the hourly wage for those overtime hours.

Example

Reginald works in Washington State and earns the state minimum wage per hour. He works 48 hour during the week. As this is 8 hours more than 40 hours, he works 8 hours of overtime. 40 hours are at the regular pay \$13.50 per hour wage and the 8 hours above 40 are at 1.5 (\$13.50) per hour (this works out to \$20.25 per hour.) His gross pay is therefore 40 × \$13.50 + 8 × 1.5 (\$13.50) = \$702.00 gross pay.

Sometimes when employees work on holidays, they can earn 100% (or more) above their pay. It all depends on the contract worked out between the company and the employees.

Practice

a) Sheila worked at the Hair Place. She is paid weekly. She worked 40 hours at \$21.50 an hour. What was Sheila's gross pay on her paycheck? __________

b) Harry is paid \$34.50 an hour and he worked 52 hours during the first week of January. What should be Harry's gross pay for the week? __________

c) Frankie got a biweekly paycheck (two weeks.) She worked 43 hours the first week then 38 hours the next week. Her hourly pay is \$42.50. What is Frankie's gross pay on her paycheck?

____

d) Henrietta worked 40 hours in the week and was paid \$580.00 gross pay. What is her hourly wage? __________

e) Jack work 42 hours in a week and was paid \$2000.00 gross pay. What was his hourly wage?

Answers:

a) 40 × \$21.50 = \$860.00

b) 40 hours at \$34.50 an hour and 12 hours at 1.5 × \$34.50 an hour, so 40 × \$34.50 + 12 × 1.5 × \$34.50 = \$2001.00

c) first week 40 hours at \$42.50 and 3 hours at 1.5 × \$42.50, second week 38 hours at \$42.50, so 40 × \$42.50 + 3 × 1.5 × \$42.50 + 38 × \$42.50 = \$3506.25

d) \$580.00/40 hr = \$14.50 per hour

e) let H = hourly wage so 40 H + (3 × 1.5) H = \$2000 or 40 H + 4.5 H = \$2000 or 44.5 H = \$2000, so H = \$44.94 per hour

7-1 Assignments

1. Figure the gross income for the following weekly pay scenarios.

	Hours	hourly pay	gross income
a)	35	$16.50/hour	____________
b)	46	$13.90/hour	____________
c)	38.5	$29.95/hour	____________
d)	49	$125.50/hour	____________

2. Delbert worked 38 hours for the week at Glorius Hair Shoppe. Delbert earned hourly pay of $22.50. What is Delbert's gross pay for the week? ____________

3. LuLu works 42 hours for the week at the Bodies'R'Us Beauty and Spa. If her hourly pay is $20.00, calculate her Gross pay ____________

4. Gloria got a weekly paycheck. She worked 47 ½ hours at a $23.50 per hour rate. What is Gloria's gross pay for the week? ____________

5. Delia had a gross pay for the week of $3449.95. If she worked 35 hours, what was Delia's hourly pay? ____________

6. Kelli was paid every two weeks. The first week she worked 44 hours and the second week she worked 41 hours. If her hourly pay was $35.50 an hour, what was her gross income for the two weeks? ____________

7. Jerry worked 49 hours and received $829.25 gross pay. What was Jerry's hourly pay?

____

7-2 DEDUCTIONS, SOCIAL SECURITY, AND MEDICARE

A DEDUCTION is money taken out of one's gross pay that goes to pay for a worker's benefits or to another institution. Deductions range from (but are not limited to) national, state, and local taxes, to union dues, to medical insurance, to social security contributions, to retirement account contributions. So deductions can be at times the money the employee will never see again as dollars in the pocket (like taxes and medical insurance) and at times the money that will be received at a later time (like retirement dollars.)

FICA DEDUCTIONS: MEDICARE and SOCIAL SECURITY

The United States of America was concerned that workers in this country would not have minimal retirement benefits when they are ready to retire at an older age. As a result, the USA developed a

MEDICARE deduction to handle health care and SOCIAL SECURITY deduction to handle income for the older citizen when they no longer work and earn a paycheck. So, by law, a little bit of every paycheck goes into a fund from which the employee will be able to have access to minimal health care and income to supplement any other resources he or she accumulates.

The combination of Medicare and Social Security is known as the FICA deduction. FICA stands for Federal Insurance Contributions Act. Typically, employers deduct FICA tax from employee paychecks and remit the money to the IRS on behalf of the employee.

Currently in 2020, MEDICARE deductions are 1.45% of all gross pay during the tax year. An additional 0.9% is deducted for the part in excess of $200,000 in a calendar year for a single filer, and in excess of $250,000 if filing jointly.

Currently in 2020 SOCIAL SECURITY is 6.2% of gross pay on the first $137,700 of gross pay during the 2020 tax year.

Example:

During the first week of 2020 Grates Worker earned gross pay of $1990.00. How much was deducted for Medicare and Social Security?

Medicare deduction is 1.45% ($1990.00) = .0145 ($1990.00) = $28.86 Social Security deduction is 6.2% ($1990.00) = .062 ($1990.00) = $123.38

Example:

During 2020 Soupy Sales earned gross pay of $2800.00 per week. This was his only income for the whole year. How much will Soupy pay in Medicare deductions and Social Security Deductions for the year.

$2800 every week will result in annual gross pay of $2800.00 (52 weeks) = $145,600.00

Since this does not exceed $200,000.00, the 1.45% Medicare is applied to his annual gross pay. 1.45% of $145,600.00 = .0145 (145,600.00) = $2111.20 annual deduction

As there is a total earning limit of $137,700.00 for the year for applying Social Security, the first $137,700 is subject to the 6.2% social security. That is .062 ($137,700.00) = $8537.40 Social Security annual deduction. The remainder of Soupy's gross pay for the year, $145,600.00 – $137,700.00 = $7900.00, will not be subject to the Social Security Deduction.

Example:

At the end of November 2020, Henry had earned gross pay for the year of $135,000.00. Henry's biweekly paycheck had gross pay of $6136.36. Figure the Medicare and Social Security deduction for this paycheck.

Adding $6136.36 to the gross pay of $135,000.00 we get $6136.36 + 135,000.00 = $141,136.36. As this is less than $200,000, the 1.45% Medicare gets applied to the $6136.36 or .0145 ($6136.36) = $88.98.

As $141,136.00 exceeds the $137,700 Social Security threshold, not all of the paycheck gross pay is subject to the 6.2% Social Security deduction. To find the part of the paycheck that takes it up to the $137,700.00 amount, we figure $137,700.00 – $135,000.00 = $2700.00. So Social Security is .062 ($2700.00) = $167.40. That means $6136.36 – $2700.00 = $3436.36 of the paycheck is not subject to the Social Security Deduction. Therefore, the Medicare deduction is $88.98 and Social Security deduction is $167.40 on this paycheck.

Other deductions taken from gross pay can include dollars deducted for things such as medical insurance, retirement contributions, union dues, or housing costs. Also deducted from gross income are tax obligations to the Federal Government, and if they apply, to State and Local government. More on this later.

Practice:

a) Find the Medicare and Social Security deductions on a paycheck the first week of 2020 with gross pay of $4400.00? ____________________

b) Wendell earned $136,600.00 gross pay through the end of October for 2020. He receives a monthly paycheck. If Wendall's gross pay is $13,660.00 on his November paycheck, how much Medicare ____________________ and how much Social Security ____________________ is deducted from his paycheck?

c) Polly worked 40 hours at $55.50 an hour. On her first weekly paycheck of 2020 had Federal Income Tax of $101.50 and had deductions for Medical Insurance of $12.50.

   What is Polly's gross income ____________________, Medicare deduction ____________________, Social Security deduction ____________________ and total deductions ____________________ for the weekly pay?

d) Here is a statement of earnings for Hardlee Worken over a one-week period. Find gross pay, Social Security, Medicare, and total deductions

Jan 12, 2020–Jan 19, 2020	**Beauty Supply Company**	
Employee: Hardlee Worken	**Position: Wharehouse staff**	
Pay Rate $19.50 / hour	**Deductions $**	
Hours: 44	**Federal Income Tax**	**134.55**
Gross pay : $_______	**Medicare Insurance _______**	**Social Security _______**
	Medical Insurance plan	**20.80**
	Retirement	**61.35**
	--	
	Total deductions	**____________**

Answers:

a) .0145 ($4400.00) = $63.80 Medicare; .062 ($4400.00) = $272.80 Social Security

b) .0145 ($13,660.00) = $198.07 Medicare; $137,700.00 – $136,600.00 = $1100.00, .062 × $1100.00 = $68.20 Social Security.

c) 40($55.50) = $2220.00 gross pay; .0145($2220.00) = $32.19 Medicare; .062 ($2220.00) = $137.64 Social Security; $101.50 + $32.19 + $137.64 + $12.50 = $283.83 total deduction

d) 40 × $19.50 + 4 × 1.5($19.50) = $897.00 gross pay for the week. .0145 ($897.00) = $13.01 Medicare, .062 × $879 = $55.61 Social Security, $ (134.55 + 13.01 + 55.61 + 20.80 + 61.35) = $285.32 total deduction so

Jan 12, 2020–Jan 19, 2020	**Beauty Supply Company**	
Employee: Hardlee Worken	**Position: Wharehouse staff**	
Pay Rate $19.50 / hour	**Deductions $**	
Hours: 44	**Federal Income Tax**	**134.55**
Gross pay: $897.00	**Medicare Insurance 13.01**	**Social Security 55.61**
	Medical Insurance plan	**20.80**
	Retirement	**61.35**
	--	
	Total Deductions	**285.32**

7-2 Assignments

Show all work!

1. Winston works at a salon and his first week's paycheck for the new year is $880.00
   How much social security is deducted from this paycheck? ________________

   How much Medicare is deducted from this paycheck? ________________

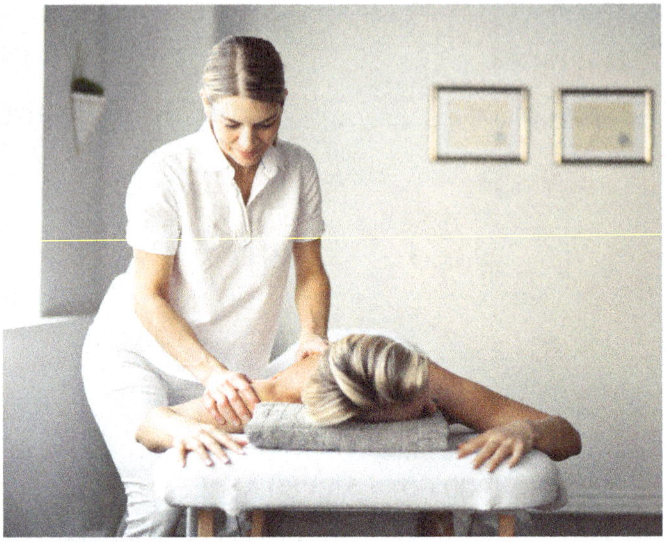

2. Fenton works at the Fancy Nail Salon, and receives $1200 in his bimonthly pay check.
   How much Social security is taken out of his check? ________________

   How much Medicare is taken out?

   ____

3. Find the Medicare and Social Security deductions on a paycheck drawn on the first week of 2020 with gross pay of $5490.00? ________________

4. Gennie's biweekly paycheck in January has a gross pay amount of $32,500.00. Figure the Medicare ________________ and the Social Security deductions ________________ for this paycheck?

5. Ronaldi earned $135,600.00 gross pay through the end of November for 2020. He is paid monthly. If Ronaldi's gross pay is $11,300.00 on his December paycheck, how much Medicare ________________ and how much Social Security ________________ is deducted from his paycheck?

6. Dolly works 40 hours a week at the Hair Care Salon and is paid $32.50 an hour. On her first weekly paycheck of 2020 had Federal Income Tax of $134.50 and had deductions for Medical Insurance of $12.50 and retirement of $25.00.
   What is Dolly's gross income ________________, Medicare deduction ________________, Social Security deduction ________________ and total Deductions ________________ for the weekly pay?

7. Here an income breaks down for Saul after a week's work at the Beauty Distributor.
   46 hours worked; $45.50 per hour pay; Federal income tax deduction $241.23; Medical insurance $15.00.

What is Saul's gross income ______________, Medicare deduction ______________, Social Security deduction ______________ and total Deductions ______________ for the weekly pay?

8. Here is a biweekly statement of earnings for Holly Deere over the two week period 1/12/20 – 1/26/20 at her job at Special Spa Experience
Find gross pay, Social Security, Medicare, and total deductions and net pay

SPECIAL SPA EXPERIENCE

Jan 12, 2020–Jan 26, 2020

Employee: Holly Deere | **Position: Beautician Assistant**

Rate $22.00 / hour

	Deductions	
Hours: 80	**Federal income Tax**	**211.20**
Gross pay : $________	**Medicare**	____
	Social Security	____
	Medical Insurance plan	**20.80**
	Retirement	**61.35**
	Union Dues	**5.78**

	Total deductions	____

Net Pay $1326.23

7-3 TAXABLE INCOME, FEDERAL INCOME TAX (FIT) DEDUCTION, AND NET PAY

A portion of one's gross pays always goes to the federal government as Federal Income Tax (FIT). This deduction is a "pay as you go "tax and is based on a workers situation laid out in a form called the W4. The W4 is used to figure TAXABLE INCOME, which is the amount of the worker's gross income that is used to figure the federal taxes.

In 2020, there are a few ways a worker can have his or her taxable income determined. On the new 2020 W4 form, a worker indicates things like marital status, filing status, number of jobs one works, prior year income info, and so on. This information creates the percentage of the gross pay

that ought to be set as taxable income. This taxable income should be in line with the worker's appropriate tax level percentage set by the government so that by year's end, the worker has had deducted from the paycheck as close to correct amount of taxes as possible. Another way is to use prior determined withholding allowances subtracted from the gross income to determine the taxable income. The taxable income determines the worker's tax from a Tax table, and that tax is deducted from the gross income on the worker's paycheck.

Once the Federal Income tax is determined, it is included as a deduction on the paycheck. This deduction is added with all other paycheck deductions, and this total deduction amount is subtracted from the gross pay for the pay period to arrive at the NET PAY. Net pay sometimes is referred to as "take-home" pay, as this is the amount the worker can take home and use as they wish.

FINDING TAXABLE INCOME USING WITHHOLDING ALLOWANCES

We will look at the second method, using withholding allowances and a percentage basis to determine 2020 taxable income.

Based on prior W4's, a worker will receive a classification based on marital status and the number of withholding allowances based on the number of dependents the worker declares. For example, a worker who is married, filing jointly with a spouse, and has three withholding allowances would be classified as M3. For someone who is single and has two withholding allowances would be classified

S2. Given the number of withholding allowances, we can calculate a dollar amount per pay period to be subtracted from the gross income to determine the taxable income for the pay period. Here is the table for the current (2020) per withholding allowances:

PAYROLL PERIOD PER WITHHOLDING ALLOWANCE 2020

Weekly: $83.00

Biweekly (every two weeks): $165.00

Semimonthly (twice a month): $179.00

Monthly: $358.00

The number of withholding allowances is multiplied by the corresponding pay period amount, then that amount is subtracted from the gross pay amount. What is left is called the **TAXABLE INCOME**.

Example:

Fabio is classified as M4. He, therefore, has four withholding allowances. He gets a gross paycheck weekly of $1100.00. The weekly allowance is $83.00, so 4 (83.00) = $332.00. Taking the withholding allowance away from the gross pay, $1100.00 – $332.00 = $768.00. Fabio's taxable income is, therefore, $768.00, which will be the amount used to figure the tax deduction for the Federal income tax.

Example:

Pinocchio receives a Semimonthly gross pay of $1875.00 He is classified S2. What is his taxable income?

He has two withholding allowances. The semimonthly allowance is $179.00. So 2 (179.00) = $358.00. Taking the withholding allowance away from the gross pay, $1875.00 – $359.00 = $1517.00. So Pinocchio's taxable income is $1517.00.

Example:

Deedle McGee has a monthly salary of $12,500 and is classified M 1. What is Deedle's taxable income?

The monthly withholding allowance for a monthly paycheck is $358.00. He has 1 withholding allowance so 1 ($358.00) = $358.00. Taking the withholding allowance away from the gross pay, $12,500.00 – $358.00 = $12142.00. Therefore, Deedle's taxable income is $12142.00.

Now, to figure the FIT deduction, we use a percentage deduction base based on one's taxable income, marital status, and one's pay period. The following tax tables break this down for weekly, biweekly, semiweekly, and monthly payments for the current 2020 tax year.

2020 PERCENTAGE METHOD TABLES FOR MANUAL PAYROLL

WEEKLY Payroll Period									
MARRIED Persons					SINGLE Persons				
If the Adjusted Taxable Wage Amount is		The amount to withhold is...	Plus this percentage...	of the amount that the wage exceeds...	If the Adjusted Taxable Wage Amount is		The amount to withhold is...	Plus this percentage...	of the amount that the wage exceeds...
at least...	But less than...				at least...	But less than...			
A	B	C	D	E	A	B	C	D	E
$0	$229	$0.00	0%	$0	$0	$73	$0.00	0%	$0
$229	$609	$0.00	10%	$229	$73	$263	$0.00	10%	$73
$609	$1,772	$38.00	12%	$609	$263	$845	$19.00	12%	$263
$1,772	$3,518	$177.56	22%	$1,772	$845	$1,718	$88.84	22%	$845
$3,518	$6,510	$561.68	24%	$3,518	$1,718	$3,213	$280.90	24%	$1,718
$6,510	$8,204	$1,279.76	32%	$6,510	$3,213	$4,061	$639.70	32%	$3,213
$8,204	$12,191	$1,821.84	35%	$8,204	$4,061	$10,042	$911.06	35%	$4,061
$12,191		$3,217.29	37%	$12,191	$10,042		$3,004.41	37%	$10,042

BIWEEKLY Payroll Period									
MARRIED Persons					SINGLE Persons				
If the Adjusted Taxable Wage Amount is		The amount to withhold is...	Plus this percentage...	of the amount that the wage exceeds...	If the Adjusted Taxable Wage Amount is		The amount to withhold is...	Plus this percentage...	of the amount that the wage exceeds...
at least...	But less than...				at least...	But less than...			
A	B	C	D	E	A	B	C	D	E
$0	$458	$0.00	0%	$0	$0	$146	$0.00	0%	$0
$458	$1,217	$0.00	10%	$458	$146	$526	$0.00	10%	$146
$1,217	$3,544	$75.90	12%	$1,217	$526	$1,689	$38.00	12%	$526
$3,544	$7,037	$355.14	22%	$3,544	$1,689	$3,436	$177.56	22%	$1,689
$7,037	$13,019	$1,123.60	24%	$7,037	$3,436	$6,427	$561.90	24%	$3,436
$13,019	$16,408	$2,559.28	32%	$13,019	$6,427	$8,121	$1,279.74	32%	$6,427
$16,408	$24,383	$3,643.76	35%	$16,408	$8,121	$20,085	$1,821.82	35%	$8,121
$24,383		$6,435.01	37%	$24,383	$20,085		$6,009.22	37%	$20,085

SEMIMONTHLY Payroll Period									
MARRIED Persons					SINGLE Persons				
If the Adjusted Taxable Wage Amount is		The amount to withhold is...	Plus this percentage...	of the amount that the wage exceeds...	If the Adjusted Taxable Wage Amount is		The amount to withhold is...	Plus this percentage...	of the amount that the wage exceeds...
at least...	But less than...				at least...	But less than...			
A	B	C	D	E	A	B	C	D	E
$0	$496	$0.00	0%	$0	$0	$158	$0.00	0%	$0
$496	$1,319	$0.00	10%	$496	$158	$570	$0.00	10%	$158
$1,319	$3,840	$82.30	12%	$1,319	$570	$1,830	$41.20	12%	$570
$3,840	$7,623	$384.82	22%	$3,840	$1,830	$3,722	$192.40	22%	$1,830
$7,623	$14,104	$1,217.08	24%	$7,623	$3,722	$6,963	$608.64	24%	$3,722
$14,104	$17,775	$2,772.52	32%	$14,104	$6,963	$8,798	$1,386.48	32%	$6,963
$17,775	$26,415	$3,947.24	35%	$17,775	$8,798	$21,758	$1,973.68	35%	$8,798
$26,415		$6,971.24	37%	$26,415	$21,758		$6,509.68	37%	$21,758

MONTHLY Payroll Period									
MARRIED Persons					SINGLE Persons				
If the Adjusted Taxable Wage Amount is		The amount to withhold is...	Plus this percentage...	of the amount that the wage exceeds...	If the Adjusted Taxable Wage Amount is		The amount to withhold is...	Plus this percentage...	of the amount that the wage exceeds...
at least...	But less than...				at least...	But less than...			
A	B	C	D	E	A	B	C	D	E
$0	$992	$0.00	0%	$0	$0	$317	$0.00	0%	$0
$992	$2,638	$0.00	10%	$992	$317	$1,140	$0.00	10%	$317
$2,638	$7,679	$164.60	12%	$2,638	$1,140	$3,660	$82.30	12%	$1,140
$7,679	$15,246	$769.52	22%	$7,679	$3,660	$7,444	$384.70	22%	$3,660
$15,246	$28,208	$2,434.26	24%	$15,246	$7,444	$13,925	$1,217.18	24%	$7,444
$28,208	$35,550	$5,545.14	32%	$28,208	$13,925	$17,596	$2,772.62	32%	$13,925
$35,550	$52,829	$7,894.58	35%	$35,550	$17,596	$43,517	$3,947.34	35%	$17,596
$52,829		$13,942.23	37%	$52,829	$43,517		$13,019.69	37%	$43,517

To calculate the tax to be deducted for the pay period means finding the

- right pay period
- the correct marital status
- the TAXABLE INCOME for that period (not the gross income)
- Calculate based on the amount of tax plus % over as indicated.

Example:

Let's go back to Fabio example, His M4 status means he has four withholding allowances. He gets a gross paycheck weekly of $1100.00. The weekly allowance is $83.00, so 4 (83.00) = $332.00. Taking the withholding allowance away from the gross pay, $1100.00 – $332.00 = $768.00. Fabio's taxable income is therefore $768.00, which will be the amount used to figure the tax deduction from the weekly Married Federal Tax Table.

On the weekly table for a married worker, want to find the line for the range of taxable income that contains $768.00. As $768 is between $609 and $1772, choose,

At least	but less than	the amount to withhold is	Plus this percentage	of the amount that the wage exceeds
$609	$1,772	$38.00	12%	$609

This means that the deduction for federal income tax owed is $38.00 plus 12% of any dollars over $609.00. To figure this overage, $768.00 – $609.00 = $159.00. Therefore, the taxable wage exceeds $609.00 by $159.00. Taking 12% of $159.00 is .12 ($159.00) = $19.08. Adding this to the $38.00, we arrive at $38.00 + $19.08 = $57.08. The FIT deduction for taxes from Fabio's gross pay is $57.08.

Example:

KelliMae works at a Hair salon and earns a biweekly gross pay of $2500. She is classified S3. What is her taxable income for this period and what will be deducted from his gross pay for Federal income tax?

S3 means three withholding allowances. The allowance for biweekly is $165.00 so

3(165.00) = $495.00. And $2500.00 – $495.00 = $2005.00. This is KelliMae's taxable income.

Using the tax table for biweekly and Single, the range income $2005.00 fits between is $1689 – $3436, so we use

At least	but less than	the amount to withhold is	Plus this percentage	of the amount that the wage exceeds
$1689	$3436	$177.56	22%	$1689

This means the deduction for federal income tax is $177.56 plus 22% of any dollars over $1689. To figure this overage, $2005.00 – $1689.00 = $316.00. Therefore, the taxable wage exceeds $1689.00 by $316.00. Taking 22% of $316.00 we get .22 ($316.00) = $69.52. Adding this to the $177.56, we get $177.56 + $69.52 = $247.08. The FIT deduction for taxes from KelliMae's gross pay is $247.08

Practice

Figure taxable income and FIT Deduction for:

a) Wilmington has gross pay of $2878 and is paid weekly. He is classified M4.
Taxable Income________________ FIT Deduction ______________________

b) Jerri's Semimonthly paycheck has a gross pay of $3450.66. She is classified as S1.
Taxable Income________________ FIT Deduction ______________________

c) Polly's monthly gross pay is $9900, and she is single with 5 withholding allowances.
Taxable Income________________ FIT Deduction ______________________

Answers:

a) $2878.00 – 4($83.00) = $2546.00 taxable income; $2546.00 – $1772.00 = $774.00; $177.56 + .22($774.00) = $347.84 taxable income.

b) $3450.66 – 1($179.00) = $3271.66 taxable income; $3271.66 – $1830.00 = $1441.66; $192.40 + .22($1441.66) = $509.57 FIT

c) $9900.00 – 5($358.00) = $8110.00 taxable income; $8110.00 – $7444.00 = $666.00; $1217.18 + .24 ($666.00) = $1377.02 FIT

PRACTICE – putting it all together

Here is pay form for Olive Oil, a Greeter at You "R" Beautiful Salon. Fill in all blanks to complete . . .

YOU "R" BEAUTIFUL SALON

Statement of earnings from February 1 to February 8, 2020

Olive Oil, Greeter	Withholding status S2
Hourly rate $16.50/hour	Deductions
Hours 45	Federal Income Tax ______________
Gross pay ______________	Social Security ______________
	Medicare ______________
	Medical Insurance $25.00
	Retirement $50.00
	--
	Total deductions ______________
Net pay ______________	

Answers: Fill in the blanks

40 hours regular pay, 5 hours overtime pay (1.5 time regular pay) is 16.50(40) = 660, 5(16.50) (1.5) = 123.75 660 + 123.75 = $783.75 Gross wage

Paid weekly, and S2 means 2 withholding allowances Weekly W/H allowance at $83.00; 2 (83.00) = $166.00 allowance $783.75 – $166.00 = $617.75 taxable income. Using table Weekly Single, find line

At least	but less than	the amount to withhold is	Plus this percentage	of the amount that the wage exceeds
$263	$845	$19	12%	$263

$617.75 – $263.00 = $354.75, $19.00 + .12 ($354.75) = $61.57 FIT; Social Security .062 ($783.57) = $48.58; Medicare, .0145 ($783.75) = $11.36; Total deductions $61.57 + $48.58 + $11.36 + $25.00 + $50.00 = $196.51; To find net pay, Gross pay – total deductions = $783.73 – $196.51 = $587.22

Or

YOU "R" BEAUTIFUL SALON

Statement of earnings from February 1 to February 8

Olive Oil, Greeter — Withholding status S2

Hourly rate $16.50/hour

Hours 45

Gross pay **$783.75**

Deduction	
Federal Income Tax	**$61.57**
Social Security	**$48.58**
Medicare	**$11.36**
Medical insurance	$25.00
Retirement	$50.00
Total deductions	**$196.51**

Net pay **$587.22**

7-3 Assignments

Show all work!!

1. BonnieSue has a gross biweekly paycheck of $3200.00. She has a W4 classification of S3. Calculate BonnieSue's Taxable income for the pay period______

   Calculate BonnieSue's Taxes for the pay period using the current 2020 tax tables ________

2. Telly has a good job at the Beauty Store who pays him in a Monthly paycheck. His gross pay for the month is $6650.00. Telly has a W4 classification of M2.
   Find Telly's Taxable income for the month ___________

   Find Telly's FIT taxes for the month ___________

3. Pollyann worked 40 hours the first week of April 2020 and made $56.00 an hour. Her W4 status is M1. What is Pollyann's taxable income for the week's paycheck ___________ and her FIT tax deduction for the paycheck? ___________

4. Xavier is a hairstylist at HairsGood Salon. He is single with three withholding allowances. His gross pay for his semimonthly paycheck was $4950.50. What is Xavier's taxable income ___________ and FIT ___________?

5. Here are details for the first monthly paycheck in 2019 for Wussel Rillson.

Gross Pay	Deductions	
$24,000.00	Federal Income Tax	$6374.19
	Social Security	$1488.00
	Medicare	$348.00
	Medical Insurance	$756.00
	Dues	$78.50

WHAT IS WUSSEL'S NET PAY FOR THE MONTH? ___________

6. Here is a weekly paystub for Ronald McWhopper.

Week of 5/4/20 through 5/11/20

Pay	Deductions	
Gross pay $3333.00	Federal Income Tax	$488.51
	Social Security	$206.65
	Medicare	$48.33
	Medical Insurance	150.00

Figure Ronald's net pay for the week. ___________

7. Here is a biweekly pay stub for Bustin Jeiber. Fill in the blanks and figure Bustin's net pay for this pay period.

Shallow Lounge

Berkley Idaho

Pay period January 1 through January 14, 2020

Employee, Bustin Jeiber, Singer/Entertainer rate, $29.50 hourly Status S2

Hours 75

Pay

Gross Pay ______________

Deductions

Federal Income Tax _________

Social Security __________

Medicare _____________

Net Pay ________________

8. Fill in the blanks and Calculate Walter White's net pay from this pay account

Albuquerque School District
JP Wynne High School
Albuquerque, NM

Week pay period: January 15, 2020 to January 22, 2020

Employee Walter White, Chemistry Teacher **Rate $45.00 hourly** **Status M4**

Pay

Hours 40

Gross pay __________

Deductions

Federal Income Tax ___________

Medicare ______________

Social Security ____________

Medical Insurance $55.00

Retirement $90.00

Dues $22.00

Net Pay ______________

7-4 MANDATORY EMPLOYER FEDERAL PAYMENTS

The Federal Insurance Contribution Acts (FICA) was set by the government to require an employee to contribute his or her Medicare and Social Security obligation. As you recall, that contribution for Medicare in 2020 is 1.45% of gross pay for the tax year (then an additional .9% on gross pay over $200,000 for the tax year if filing separately, over $250,000 if filing jointly) and for Social Security is 6.2% of the first $137,700 of gross pay for the tax year. The worker's employer is responsible for sending that contribution to the Federal Government on behalf of the worker.

FICA also requires an employer to match all the employee's Medicare and Social Security contributions and send that as well to the Federal Government. Let's say Joe the Worker has $4,567 in Medicare and Social Security contributions for the year. Joe's employer must also send a matching $4,567 to the Federal Government, according to the FICA rules.

For example: Grate Bodie owns a small health and beauty company. He has three employees and the company payroll register for one-week lists the following:

Employee	Status	Gross	Medicare	Soc. Sec.	FIT	Net
Goodie Worker	M1	1945.00	28.20	120.59	197.36	1598.61
Neve R. Giveups	M3	2030.00	29.44	125.86	179.54	1695.16
Al Outperformer	S1	1085.25	15.74	67.29	66.19	936.03

What is Grate's total FICA tax obligation for these three employees? In other words, how much does Grate send to the Federal Government?

We add up all Medicare and Social Security amounts,

so $28.20 + $29.44 + $15.74 + $120.59 + $125.86 + $67.29 = $387.12 FICA Tax.

And Grate must match this total, so double the FICA tax is 2 ($387.12) = $774.24. So Grate sends in a total of $774.24 to the Federal Government for the week.

FEDERAL UNEMPLOYMENT TAX ACT (FUTA)

Another obligation of an employer is the result of the Federal Unemployment Tax Act, or FUTA. Employers, not employees, are required to pay the unemployment tax and is not a deduction from an employee's pay.

As of 2020, the current FUTA rate is 6% of the first $7000 of an employee's gross pay for the year. However, each state government is the body responsible for paying out unemployment benefits to assist workers who have unexpectantly been laid off. So an employer needs to meet a state unemployment tax obligation as well under their respective State Unemployment Tax Act, or SUTA.

By meeting the individual state unemployment tax obligation, a credit of up to 5.4% of gross pay (up to $7000 an employee) is applied to the FUTA.

As individual states' unemployment tax rates vary it is not practical to figure every state possibility. But we can say if unemployment obligations are met and on time, we can assume the maximum credit on FUTA. As 6% – 5.4% = .6%, the then the FUTA tax burden is .6% of an employee's gross pay. And .6% of $7000 = $42.00, the maximum amount needed an employer needs to send in for an employee.

Example:

Ronald McDonald gross pay is $4500.00 for three weeks of work. If the 5.4% credit is in play for FUTA, what is Ronald's employer FUTA obligation for this period?

FUTA is .6% of the $4500.00 or .006($4500.00) = $27.00. His employer sends $27.00 to the federal Government.

Example:

Carny Vail works at the GreatFun amusement park. By August, his accumulated gross pay for the year was $14,010. If the 5.4% credit is in play for FUTA, what is GeatFun's FUTA obligation through August?

As there is more than $7000.00 in gross pay, his employer's obligation is only .6% on the first $7000.00, or .006 (7000.00) = $42.00.

Practice

a) The first paycheck for Harry in 2020 had a gross pay of $5650. What is Harry's employer FICA total payment to the Federal Government as a result? ______________

b) Gully's year to date gross pay in 2020 is $3400. What is Gully's employer's FUTA obligation to the Federal Government as to Gully's pay to date (assume .6% rate)? ___________

c) Joanie's year to date gross pay in 2020 is $11,700. What is Joanie's employer's FUTA obligation to the Federal Government as to Joanie's pay (assume .6% rate)? ___________

d) For the following employee's, figure their Medicare and Social Security deductions, the matching contributions by the employer, the total amount FICA the employer send into the federal government and the total amount of FUTA payment (assuming .6% rate)?

EMPLOYEE	2020 TO DATE EARNINGS $	CURRENT PAY CHECK GROSS $	MEDICARE DEDUCTION $	SOCIAL SECURITY DEDUCTION $	EMPLOYER MEDICARE MATCH $	EMPLOYER SOCIAL SECURITY MATCH $
Ross, B	65,000	6000				
Dorn, A	76,000	7000				
Bella, C	4500	340				
Vann, V	6900	600				

The FICA payment made by the employer = ________________

The FUTA payment made by the employer = ______________

Answers:

a) .0145 ($5650.00) = $81.93 Medicare; .062 ($5650.00) = $350.30 Social Security; 2 ($81.93 + $350.30) = $864.46

b) .006 ($3400.00) = $20.40

c) tax on only first $7000.00, .006 ($7000.00) = $42.00

d)

EMPLOYEE	2020 TO DATE EARNINGS $	CURRENT PAY CHECK GROSS $	MEDICARE DEDUCTION $	SOCIAL SECURITY DEDUCTION $	EMPLOYER MEDICARE MATCH $	EMPLOYER SOCIAL SECURITY MATCH $
Ross, B	65,000	6000	87	372	87	372
Dorn, A	76,000	7000	101.50	434	101.50	434
Bella, C	4500	340	4.93	21.08	4.93	21.08
Vann, V	6900	600	8.70	37.20	8.70	37.20

FICA payment = 2 (87 + 101.50 + 4.93 + 8.70 + 372 + 434 + 21.08 + 37.20) = $2132.82. For FUTA, Ross and Dorn both have exceeded $7000 gross YTD, so no FUTA, Bella full $340 subject to FUTA, so .006 (340) =$2.04 , Vann, only $100 subject to FUTA, so .006 (100) = $.60, so FUTA payment = $2.04 + $.60 = $2.64.

7-4 Assignments

1. On Delia's January 2020 paycheck, her Medicare deduction is $45.50 and her Social Security deduction is $131.55. What should Delia's employer send in for the FICA payment?

2. On his paycheck for January 2020, Bing had gross earnings of $5400. What is his employer's FICA payment to the Federal Government based on his gross pay? _______________

3. By the end of November 2020, Geraldine has accumulated gross income of $136,000.00 in 2020. Geraldine's December paycheck was for $12,000.00. What should Geraldine's employer send into the Federal Government to cover the FICA payment for this last paycheck? _______________

4. Jimmy's year to date gross pay in 2020 is $5400.00. What is Jimmy's employer's FUTA obligation to the Federal Government as to Jimmy's pay to date? (assume .6% rate) _______________

5. Ken Doll's year to date gross pay in 2020 is $7950.00. What is Ken's employer's FUTA obligation to the Federal Government as to Ken's year to date pay? (assume .6% rate) ______________

6. AshleySue Year to date gross pay for 2020 is $6600.00. Her next paycheck had a gross pay of $1200.00. For this paycheck, what does AshleySue's employer needs to send in for the FUTA obligation? (assuming the .6% rate) ______________

7. For the following employee's paychecks, figure their Medicare and Social Security deductions, the matching contributions by the employer, the total amount FICA the employer send into the federal government, and the total FUTA payment for the current paycheck? (assuming .6% rate)

EMPLOYEE	2020 TO DATE EARNINGS $	CURRENT PAY CHECK GROSS $	MEDICARE DEDUCTION $	SOCIAL SECURITY DEDUCTION $	EMPLOYER MEDICARE MATCH $	EMPLOYER SOCIAL SECURITY MATCH $
Bossie, M	4000	1000				
Verne, J	8700	2000				
Cella, D	4300	1400				
Finn , F	9000	3500				

The FICA payment made by the employer = ______________

The FUTA payment made by the employer = ______________

Practice Tests

Answer key in the back of the book

Practice Test Number 1 Covers Chapters 1, 2, and 3

Use the operations indicated and simplify the following

1. $\frac{4}{7} \div \frac{5}{8}$ __________

2. $\frac{11}{42} \cdot \frac{6}{55}$ __________

3. $5\frac{2}{3} + 11\frac{3}{7}$ __________

4. $\frac{6}{19} + \frac{3}{38}$ __________

5. $6^2 \div 4 + 5 - 3 \cdot 2 + \frac{8+4 \cdot 2}{8}$ __________

6. $15 - 16 \div 2 \cdot (1 + 6^2 \div 4 \cdot 2 - 17) + 4$ __________

7. $2\frac{2}{7} \cdot 3\frac{1}{9}$ __________

8. $\frac{8}{9} - \frac{3}{5}$ __________

9. $25\frac{4}{5} - 8\frac{1}{5}$ __________

10. $\left(7\frac{2}{3}\right)\left(2\frac{1}{11}\right)$ __________

11. $6\frac{2}{3} \div 1\frac{1}{4}$ __________

12. $\frac{5}{7}$ of 560 __________

13. $\sqrt{676}$ __________

14. $\sqrt{2459}$ to the nearest hundredth place ______________

15. $\sqrt[3]{343}$ ______________

16. $\sqrt[3]{23.15}$ (to nearest hundredth place) ______________

17. $\sqrt[11]{815}$ (to the nearest hundredth place) ______________

18. $[(1 + .08)^4 - 1]$ (to nearest hundredth place) ________________

19. $\dfrac{(5)(9)-5(6-4)^3}{(5-2)^2-2^2}$ ____________________

20. Wendy hiked an average of 14 miles per day for five days. Peter hiked an average of $5\frac{2}{5}$ miles per day for the five Days. Pan hiked 3/5th of the total miles Wendy and Peter hiked combined. What were the totals of each for the five days? Wendy ______________

Peter ________________

Pan ________________

21. Solve for y $\dfrac{y}{8} = \dfrac{4}{13}$ ______________

22. Solve for x $\dfrac{4}{11} = \dfrac{20}{3x}$ ____________________

23. Solve for x given $3 + \dfrac{x}{6} = 5$ __________________

24. Solve for x $\dfrac{5}{2x} + \dfrac{7}{7x} = 12$ ______________

25. Solve for y $\dfrac{3y}{8} - \dfrac{3}{13} = \dfrac{11y}{13} + \dfrac{1}{8}$ ______________

26. Solve for x, $\dfrac{x+4}{3} = \dfrac{2x-5}{2}$ _________

27. For the formula $y = 3x + 7$ Solve for x, given $y = 2$ ____________

28. Solve for P in $I = Prt$ given $I = 350$ $r = .08$ $t = 4$ _____________

29. For the formula $\dfrac{r+3}{2p-4} = \dfrac{r-5}{p}$ solve for r given $p = 10$ ____________

30. Solve for Y given $610 + .2Y = Y$ __________________

31. Simplify $2x^2 + 11x + 8 - 2x^2 + 12 - 2x$ ___________________

32. Solve for w given $8(20 - w) + 15w = 20$ ________________

33. Calculate for A in $A = m[(1+rt)^n - 1]$ given $m = 100$, $r = .006$, $t = \dfrac{1}{12}$ and $n = 180$ (to the nearest hundredth place) _______________

34. Chef Boyardee had 9 pizzas to every 4 calzones in his restaurant. If he had 108 pizzas, how many calzones did Chef have? __________

35. Fab Ulousweathy had 25000 Canadian dollars. He crosses the border for for a whirlwind USA vacation. He exchanges his Canadian dollars to US dollars. The exchange rate is 1.23 Canadian dollars for $1 dollar (US). How much in US Dollars can Fab receive? ____________________

36. Speedy Joe drives his motorcycle at an average of 72 miles per hour. How many Kilometers does he travel in 180 minutes, given that there are 1.61 Kilometers to 1 mile? ____________

37. For the formula $\frac{8\text{ wt}}{5\text{ m}} = 32\text{ mx}$ solve the formula for w. ______________

38. For the formula $3y = \frac{3p+t}{5yr^2}$ solve the formula for t __________

39. **a)** 45% of what number is 58.5? ________

b) 31.56 is what % of 789? ________

c) 120% of 34500 is what number? ________

40. **a)** 950 students of the 3500 total students at ReallySmart University are in the Junior class . What % of the total student body are Juniors? ________

b) 35.4% of the Paintings in the Wilson Museum were by an artist Hernando. If there were 23 Hernando paintings, how many paintings total were there in the Wilson Museum. ________

c) There were 550 Reality TV shows on the major TV Networks in 2019. 84% of these shows dealt with competition of some sort. How many were competition shows? ________

41. Convert 150 kg = ____ lb ____________

42. Convert 21,500 mg = ____ g ____________

43. Convert 500 fluid ounce to Liters ____________

44. Convert 50 feet per minute to meters per second ____________

45. Convert 15,000 mL = ______ kL ________________

46. Convert 150 pt / 1 hour to______ qt / 1 min ________________

47. What is the longer length? 4.52 miles or 7.18 Kilometers? __________

48. What is a bigger volume? 56.89 pts or 7.03 Gallons ____________

49. $\frac{1}{100}$ km = ______ meters ____________

50. 3,456 mL = _____ cL ____________

51. 18 gallons = _____ cubic inches ________

52. Barney bought 4 cartons of shampoo. Each carton contains fifty 22.5 fluid ounce bottles. How many Quarts of shampoo did Barney purchase? _________

53. Fabio Hair sells conditioner at $18.50 per 600 mL. What is the price per fluid ounce? _________

54. Sir Hair-a-lot bought a barber chair that weighed 16,500 ounces. What is the chair's weight in pounds? ________

55. Freddie has 6 Liters of Liquid Beverage that is 20% strawberry juice. How many liters of an 80% strawberry juice should be added to the the 6 Liters of beverage to make it 75% strawberry juice? _________

56. Helene's Hair Emporium sponsored a Mother/Daughter Beauty Show where they sold tickets. The price for Adult tickets was $10.00 and the price for Children's tickets was $6.50. If they sold 200 tickets in all and brought in $1,720 in sales, how many of adult and how many of children tickets were sold? ________ ___________

57. Gul Yibbons wants to make 20 pounds of trail mix using nuts and chocolate chips. His budget requires that the trail mix costs him $7.60 per pound. Nuts cost $9.00 per pound and chocolate chips cost $2.00 per pound. How many pounds of nuts and how many pounds of chocolate chips should he use? ____________ ______________

58. Aretha has $40,000 to invest and hopes to earn 7.1% interest per year on her investment. She will put some of the money into a stock fund that earns 8% per year and the rest into bonds that earns 3% per year. How much money should he put into each fund? _________ ________________

Practice Test Number 2, covers Chapters 4 and 5

1. Convert the following Customary times to Military times

a) 4:30 AM ______

b) 5:25 PM ________

c) 11:59 AM ________

2. Convert the following Military times to Customary time

a) 2132 ________

b) 0344 _______

c) 1145 ____________

3. Giving the hours in simplified mixed number fraction, What is the total number of hours of appointments schedules from

Haircut $1\frac{3}{8}$ hours; Shampoo $\frac{4}{7}$ hour ; Styling 2 hours ; Perm wave 3 ½ hours; Facial $1\frac{7}{8}$ hours; Scalp care ½ hour _______________

4. On Tuesday Winnie worked from 7:56 Am to 11:32 AM. What is the total time in hours and minutes Winnie worked that morning. ________

5. Francee worked from 10:55AM to 6:11PM straight without a break. How many hours and minutes did Francee work? ___________

6. Horatio wanted to check the number of hours worked. Find the total time in hours and minutes if he worked from 9:15 AM to 11:48AM and from 12:44PM to 5:06 PM _________

7. For the following time schedule of activities by a cosmetology student, find the time in hours and minutes of activities.

Date	Lecture	Hours	Demonstration	Hours	Practice	Hours
10-Jul	haircutting	2 ½	haircutting	1 ¾	haircutting	2
11-Jul	haircutting	1 ½	haircutting	1 ¼	haircutting	1 ¾
12-Jul	shampooing	1	shampooing	1	shampooing	1
13-Jul	Business principles	1 ¼				
14-Jul	Styling	1 ¼	styling	1	styling	1 ½

8. During a 9-hour period, Stacielou scheduled the following appointments at her beauty shop. 1.5 hour haircut; 1.75 hour Styling; 2.6 hour Massage; and 1.2 hour haircut. How much time in hours and minutes did Stacielou have for lunch and/or breaks? ___________

9. Terry worked 115 minutes on inventory. She then worked $4\frac{3}{4}$ hours on store decoration. How much time did she work altogether? Give answer in combined hour and minutes.

10. Sonny needs to schedule customers for sun tanning. If each session takes 40 minutes, how many tannings can Sonny schedule during a $6\frac{1}{2}$ hour work shift? Moreover, if she has any time left over, how much time is it? _________ ______

11. Polly had $5\frac{5}{6}$ hours scheduled for appointments at her salon. She got word that two customers had to cancel, one with a 55-minute appointment and another with a 100-minute appointment. After this, how much time was used for appointments? Show answer in hours and minutes.

12. Topher purchase cartons of shampoo for the supplier his shop. The total list price for his order is $540. The supplier offers an 8% discount to Topher. How much does Topher send to the supplier for this order? ___________

13. Petunia bought a massage table listed at $2700 by the manufacturer. The manufacturer offers a 6.8% discount on their price. How much did the table cost Petunia? ___________

14. Wonder Hair Store bought goods at listed price of $4,300 from the Beauty Distributors. Beauty Distributors offered Wonder Hair 20/5 trade discount, which is a 20% discount followed by a 5% discount. How much does Wonder Hair pay Beauty Distributor for this order? ___________

15. Federico's, a nail salon, bought nail products from National Beauty Distributors with a trade discounts of 25/10/5, which is a 25% discount , followed by 10% discount, followed by 5% discount. If the list price of the products added up to $750.00, what was the actual cost to Federico's. ___________

16. Debbie's Delighful HairDo's sells haircuts for $25.00 each. Debbie offers a 35% off special. What is the special price to the customer? ________

17. A 20/10 trade discount is equal to what single percent discount? __________

18. A 30/5/5 trade discount is what single discount equivalent? ________

19. Barry's Beauty and Spa sells massage oils at $23.50 a bottle. If there is a 6.7% sales tax on each sale and a customer buys two bottles, how much does the customer pay? ___________

20. Kelli sells wash cloths at her beauty store. She buys them from her distributor for $23.00 a set, and then marks up the cloths 85%. What is the selling price for a set of wash cloths? ___________

21. Bath and Beauty Store bought conditioner at $12.00 a bottle from the distributor. They marked up the cost 55%. What is the selling price of the conditioner in their store? _________

22. LovelyLocks Hair Salon bought combs at $2.50 per comb from the All World Distributor. Lovely Locks marked up the cost 35%, then to sell more of them later, they marked down the price 25%. What was the final sales price per comb? ____________

23. A professional equipment store bought a beauty Chair listed a $1,200 from the manufacturer who offered a 10% trade discount. The chair was marked up 20% by the buyer to sell, then after a few weeks that price was marked up another 10%. What is the final selling price for the chair?

____

24. Janie's, a beauty shop, originally sells Mineral oils at $32.50 a set in January. Janie marks up the oils 45% in February, then marks them down 24% in March. An 11.2% sales tax is set on every purchase. What does a customer pay for a set bought in March? ______

25. In August, Radiant Living Spa showed a total revenue for services and sales of $17,567.70. The Spa total costs and expenses for August was $10,042.90. What, if any, was Radiant Living Spa's profit for the month of August? __________

26. Betsy's Beauty Shoppe had a profit of $89,000.00 for 2019. If Betsy's total sales and revenue for 2019 was $104,000.32, what was Betsy's 2019 cost and expenses? __________

27. ReeAnne charges $25.00 a haircut. If during a week she did 18 haircuts and had out of pocket expenses of $115.00 for the week, what was ReeAnne's profit for the week? __________

28. Nell's Beauty and Healthy Living store purchased 500 jars of moisturizer from wholesaler at a cost of $6250.00. Nell wants to make a profit of $4.50 on each jar she sells at her store. If Nell sells all 500 bottles, how much does Nell need to price a jar and how much in total sales dollars is needed to achieve her goal? ____________ ____________

29. On Wednesday, Daria charged $45.00 each for the facials she did for customers. Daria's total expenses for her supplies for Wednesday were $27.00, and were her only expenses. After taking out her cost of supplies, Daria earned $ 378.00 (profit) for the day. How many facials did Daria do on Wednesday? __________

30. During December, Rolf did 48 hair stylings in his hair salon. This was his total workload for December. He charged a regular price for the first 30 hair stylings, and then offered a discounted price of 40% off the regular price for the remaining 18 hair services. His total cost for supplies and expenses for December was $900.00.

What was Rolf's regular charge per service and what was the discounted charge per service if he brought in $1303.20 in profit for the month of December?

31. Here was a ledger of costs and sales for Doris's business for the month of June.

Cost, June	
Chair Rental	$450
Supplies	$700
Equipment	$150
Taxes	$25
License	$10
Miscellaneous	$65

Revenue, June
30 Haircuts @$20 per haircut
25 Pedicures @$18.50 per pedicure
10 Facials @$42 per facials
10 Hair coloring @$25 per hair coloring
16 Manicures @$21 per manicure
$410 in tips

What was Yvette's profit for June? __________

32. A researcher for an insurance company wants to collect data to see if color of one's car is related to likelihood of an accident. The researcher looks at recent police reports on total-loss collision and finds the following: out of 100 reports, 20 were from blue cars, 24 from green cars, 33 from red cars, 2 from white cars, 16 from black cars, and 5 from silver cars.

Create a frequency table and a bar graph for this data.

33. For the data in assignment 32, figure the degree of the central angle for each pie section, and create a Pie Graph.

34. The following data represents the number of minutes pedicures took on a Tuesday Fannie's Fabulous Spa.

35	36	45	55	76	88	44	57	62
56	57	59	76	82	77	68	65	62
65	64	72	81	66	63	45	67	66

Create a frequency table, with time interval for 10 minutes, starting with 30 minutes, ending with 90 minutes, then create a histogram from the table.

35. For this histogram representing scores made by students on a cosmetologist standards test, answer

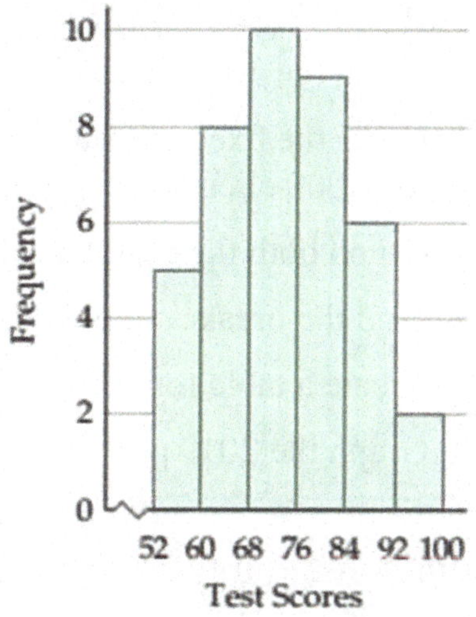

a) How many students scored between 68 and 92 on the test? ________

b) How many students total took the test? ________

c) What was the percentage of students, (to the nearest 10th of a percent) who scored less than 68? ________

36. JennieLynn sells scented candles from a kiosk. She sells them at $35.00 a candle. Her cost for the candles is $21.00 a candle. She has fixed costs on Monday for her kiosk of $280.00.

a) Do a cost table for total cost dollars and number of candles sold for Monday. In your table, do intervals of five candles sold up to 60.

b) Do a sales table for sales dollars and number of candles sold for Monday. In your table, do intervals of five candles sold up to 60.

c) On the same graph, find a line graph for costs and for sales, with number of candles sold as the horizontal axis and dollars cost/sales as the vertical axis.

d) Find the breakeven point, where sales and costs are the same for a given number of candles sold.

37. Susie's Soothing Spa generated revenue for 2019 as follows:

HEALTH SERVICES OR PRODUCTS	Massages	Body Treatments	Pedicure and Manicures	Facials	Retail sales
2019 Revenue	$26,000	$15,000	$12,000	$30,000	$10,000

Susie also calculated costs for 2019 as follows

Fixed costs – those costs that are independent of number of clients

Fixed cost type	Rent	utilities	taxes	licenses	advertising	miscellaneous
Fixed cost $	20,000	3,000	1,500	600	400	1,500

Variable costs – those costs impacted by the number of clients

Variable cost types	labor	store products	supplies	miscellaneous
Variable cost $	31,000	8,500	9,000	4,000

Susie has calculated 1,500 clients were served in 2019.

Using the data, Find the

- average of 2019 sales/revenue per client.
- 2019 average variable cost per client.
- Using the average sales/revenue per client, create a table of sales per client (0–4,000 clients).
- Using the fixed costs and average variable cost per client, create a table of total cost per client (0–4,000 clients).
- Graph both the cost data and sales data on the same graph.
- Find the breakeven points, where the number of clients has the same sales and cost.
- Create a table for 2019 profit.
- Graph the 2019 profit data per client (0–4,000 clients).

Practice Test Number 3, covers Chapters 6 and 7

1. If on a simple loan, I = interest, P = Principal, R = interest rate, and T = time What formula would equal P?

   a) $P = I R T$ b) $P = \frac{I}{R T}$ c) $P = \frac{R}{I T}$ d) $P = \frac{T}{I R}$

2. If on a simple loan, I = interest, P = Principal, R = interest rate, and T = time What formula would equal R?

   a) $R = I P T$ b) $R = \frac{I}{P T}$ c) $R = \frac{P}{I T}$ d) $R = \frac{T}{I P}$

3. How much simple interest is there on a $5,000 loan at 9.9% interest? ___________

4. What is the interest % rate on a Simple loan principal of $13,400, if the interest is $2,310?

   ____

5. BetsySue took out a $7,000 simple loan from the Wegotthebucks Bank at 7.9% interest and paid it off at the end of the year. What did BetsySue give the bank to settle his loan at the end of the year? _____________

6. If KellyAnne had $1,100 interest on a simple loan for a car at 3.5% interest after one year, How much did KellyAnne originally borrowed? ______________

7. Winning Hair Shoppe took out a business loan for upgrades in the store. If the loan amount was $6,000, and a simple annual interest of 5.5%.

   a) After 3 ½ years, what does winning pay back? __________

   b) How much interest is paid? __________

8. Nelson borrowed $3,000 from MoneyMoneyMoney Bank for 75 days at annual simple interest of 9.5%. How much did Nelson pay the bank on the 75th day? ____________

9. Bobbikens' Beauty purchased $80,000 in a bond at a simple 4.68 % annual interest charged monthly with a maturity of 5 years. How much does Bobbiken's have in the investment after maturing in 5 years. __________

10. Bethanie's Beauty Shoppe borrowed $5,000 from a bank at 6% annual interest. Bethanie's paid back the loan on the 595th day. How much is paid back? ____________

11. A $35,000 investment at 7.55% annual interest is compounded annually for 26 years.

    a) What is the ending balance of the investment? ______

    b) How much interest is earned? __________

12. Dilbert invests $15,000 for 6 years at a bank. The annual interest on the loan is 4.6% compounded monthly.

    a) What is in his account balance after 6 years? __________

    b) How much interest does Dilbert earn on his investment? _________

13. Molly borrowed money to repair her store, The HairAffair. She borrows $10,500 from a bank at an annual interest of 8.5% compounded quarterly. She pays the loan back after 3 years. What is her payback amount after the 3 years? _________

14. Sabrina invested $12,000 into an account earning 6.7% annual interest compounded weekly for 8 years. What is in the account after the 8 years? ________________________

15. Loretta wants to have an account worth $1,800,000 after 20 years. If Loretta has a monthly compounding account at 5.55 % annual interest, what single principal deposit does she need to make to accomplish this? ______________

16. George Jetson deposited a lump sum of money in a daily compounding account at 19% annual interest for 900 days. If there was $4,450 in the account after the 900 days, how much did George initially deposit? ___________

17. Norman invests in a 6.7% annual interest investment for 10 years that is compounded annually. His goal is $2,800,000 total at the end, so what does Norman need to deposit at the start of the account? __________

18. Bartholomew takes out an annuity where $4,400 is deposited at the beginning of each year for 25 years. If the annual interest rate is 4.5% compounded annually, what is the balance after the 25th deposit? ______________

19. An annuity is set up where $230 is deposited at the beginning of each week for 5 years. If the annual interest rate is 11.5 % compounded weekly, what is the balance after the last deposit?

____

20. MaryAnne invests in a monthly annuity where $1,500 of her paycheck is deposited at the beginning of each month for 30 years. If the annual interest rate is 2.99% compounded monthly, what would MaryAnne have in the account after the last deposit? ________________

21. GerriSue wants to purchase a beauty franchise. The sales price is $29,000.00, and she finds a 5.5% annual interest loan that is compounded monthly, to be paid in monthly payments for 7 years. What would GerriSue's monthly payment be? _________

22. Kelso wants to buy a new home that sells at $220,000. He pays $20,000 up front, and then finances the rest at 3.5% annual interest for 30 years, to be paid quarterly. If his loan is compounded quarterly, what would his quarterly payments be? _____________

23. Biff made a daily deposit in an annuity account that accumulated $11,000 after 6.5 years. If the account is compounded daily at 15.2% annual interest, what was Biff's daily deposit into the account? _________________

24. Fred had a Gross pay for the week of $2,000, out of which taxes were deducted at $214.00, Social Security at $124.00 and Medicare at $29. Assuming no other deductions, what was Fred's Net pay for the week? ____________

For 25–27, assume that any hours of work over 40 in a week is considered overtime pay.

25. Jennie works 38 hours last week at the Beauty Bar. Her hourly wage is $32.50. What was Jennie's gross income for the week? ______________

26. Naomi works 45 hours for the week at the TopHat Hair Shop. If her hourly pay is $19.50, calculate her gross pay for the week. ____________

27. Hank is paid biweekly for his job at the HairShoppe. For the current paycheck, he worked 43 hours the first week, then 40 hours the second week. He is paid $43.50 an hour. What is his gross pay on the biweekly paycheck? ________________

28. Kelli works at a salon and her first week's paycheck for the year, 2020, is $1,350.00

a) How much is deducted for Social Security from this paycheck? _____________

b) How much is deducted for Medicare from the paycheck? _______________

29. In 2020, Delores has accumulated a gross pay amount of $135,800. Her next paycheck was for a gross pay amount of $3,200.

a) How much is deducted for Social Security from this paycheck? _____________

b) How much is deducted for Medicare from the paycheck? _______________

30. Fiona has a gross weekly paycheck of $2,700 the first week of 2020. She uses withholding allowances to figure her taxable income for Federal Income Tax (FIT). She has a W4 classification of S2.

a) Calculate Fiona's Taxable income for the pay period______

b) Calculate Fiona's Taxes for the pay period using the current 2020 tax tables___________

31. In 2020 Clark had a January gross monthly paycheck of $11,500. He uses withholding allowances to figure her taxable income for Federal Income Tax (FIT). He has a W4 classification of S2.

a) Find Clark's taxable income for this paycheck ___________

b) Find Clark's federal taxes for this pay period using the 2020 tax tables ___________

c) Find Clarks Social Security deduction _________

d) Find Clark's Medicare deduction _________

32. Using the 2020 fill in the blanks and calculate Vincent Monet's net pay from this biweekly pay account. Use the current 2020 tax tables

Michelangelo Art Supply

333 Picture Perfect Place

New York, New York 14567

Two Week pay period: January 14, 2020 to January 28, 2020

Employee Vincent Monet, Painter	**Rate $25.00 hourly**	**Status**	**M3**
Pay	**Deductions**		
Hours 80	**Federal Income Tax** ___________		
	Medicare _____________		
Gross pay __________	**Social Security** ___________		
	Medical Insurance $111.00		
	Dues $40.00		
Net Pay ______________			

33. For the employee's at All Beauty Distributors, given their respective gross pay for the first week of 2020 and their W2 status and using the 2020 tax tables, find the taxable income, the FIT, Social Security, Medicare deduction, and the Net pay (assume no other deductions.)

Name	Week's Gross pay	W2 Status	Taxable Income	Federal Income Tax	Social Security	Medicare	Net Pay
Shelly	$2,150	S2					
Sylvia	$11,456	M3					
Doug	$3,510	S1					
Ellie	$600	M2					

34. Taylor worked at AllCare Health and Beauty and earned gross pay of $5,000 to date in 2020. Her next paycheck was for $1,300.

a) Find Taylor's Social Security and Medicare deductions for this paycheck. ______

b) Calculating Taylor's employer's FICA obligation based on Taylor's paycheck, how much in total does the employer send to the federal government? __________

c) What is Taylor's employer's FUTA payment based on Taylor's paycheck. (Assume the .6% rate applies). _________

35. NormaRae worked at a Hair Salon and earned gross pay of $6,650 to date in 2020. Her next paycheck was for $1,150.

a) Find NormaRae's Social Security and Medicare deductions for this paycheck. ______

b) Calculating NormaRae's employer's FICA obligation based on NormaRae's paycheck, how much in total does the employer send to the federal government? __________

c) What is NormaRae's employer's FUTA payment based on NormaRae's pay (assume the .6% rate applies). _________

36. BillyBob worked at a Spa and earned gross pay of $7,600 to date in 2020. His next pay check was for $950.

a) Find BillyBob's Social Security and Medicare deductions for this paycheck. ______

b) Calculating BillyBob's employer's FICA obligation based on BillyBob's pay, how much in total does the employer send to the federal government? __________

c) What is BillyBob's employer's FUTA payment based on BillyBob's pay (assume the .6% rate applies). _________

Chapter Assignment Answers

Answers for 1-1 Assignments

1. a) APPROXIMATE
   b) APPROXIMATE
   c) EXACT
   d) EXACT
   e) APPROXIMATE
   f) EXACT
   g) EXACT
   h) APPROXIMATE

2. a) FALSE
   b) TRUE
   c) FALSE
   d) TRUE
   e) TRUE

Answers for 1-2 Assignments

1. $\frac{8}{16}=\frac{1}{2}$

2. $\frac{14}{18}=\frac{7}{9}$

3. $\frac{28}{35}=\frac{4}{5}$

4. $\frac{145}{150}=\frac{29}{30}$

5. $\frac{54}{108}=\frac{1}{2}$

6. $\frac{51}{63}=\frac{17}{21}$

7. $\frac{924}{1016}=\frac{231}{254}$

8. $\frac{4545}{9999}=\frac{505}{1111}$

9. $\frac{1}{9}+\frac{5}{18}=\frac{7}{18}$

10. $\frac{1}{9}+\frac{7}{20}=\frac{83}{180}$

11. $\frac{3}{7}+\frac{1}{15}=\frac{52}{105}$

12. $\frac{1}{12}+\frac{1}{9}=\frac{7}{36}$

13. $\frac{7}{8}+\frac{6}{7}=1\frac{41}{56}$

14. $\frac{11}{20}+\frac{11}{40}=\frac{33}{40}$

15. $\frac{2}{5}-\frac{1}{5}=\frac{1}{5}$

16. $\frac{2}{5}-\frac{5}{16}=\frac{7}{80}$

17. $\frac{3}{4}-\frac{2}{15}=\frac{37}{60}$

18. $\frac{7}{8}-\frac{7}{16}=\frac{7}{16}$

19. $\frac{1}{2}-\frac{3}{16}=\frac{5}{16}$

20. $\frac{9}{20}-\frac{1}{7}=\frac{43}{140}$

21. $\frac{1}{b}+\frac{5}{a}=\frac{a+5b}{ab}$

22. $\frac{1}{xy}+\frac{7}{2w}=\frac{2w+7xy}{2xyw}$

23. $\frac{3t}{7}+\frac{5t}{14}=\frac{11t}{14}$

24. $\frac{t}{p}-\frac{y}{x}=\frac{tx-py}{px}$

25. $\frac{ab}{cd}+\frac{ef}{gh}=\frac{abgh+cdef}{cdgh}$

26. $\frac{1}{f}+\frac{h}{4}=\frac{4+fh}{4f}$

27. $2\frac{1}{7}+9\frac{11}{15}=11\frac{92}{105}$

28. $4\frac{14}{15}+1\frac{7}{12}=6\frac{31}{60}$

29. $7\frac{3}{10}+7\frac{1}{16}=14\frac{29}{80}$

30. $7\frac{1}{7}+1\frac{2}{9}=8\frac{23}{63}$

31. $5\frac{4}{9}+8\frac{3}{7}=13\frac{55}{63}$

32. $9\frac{8}{9}+6\frac{4}{7}=16\frac{29}{63}$

33. $8\frac{5}{7}-2\frac{2}{15}=6\frac{61}{105}$

34. $9\frac{8}{9}-6\frac{1}{2}=3\frac{7}{18}$

35. $10\frac{1}{12}-7\frac{3}{4}=2\frac{1}{3}$

36. $5\frac{2}{5}-1\frac{1}{2}=3\frac{9}{10}$

37. $8\frac{7}{20}-4\frac{1}{4}=4\frac{1}{10}$

38. $7\frac{5}{9}-6\frac{3}{4}=\frac{29}{36}$

Answers for 1-3 Assignments

1. $\frac{1}{9}\cdot\frac{5}{7}=\frac{5}{63}$

2. $\frac{4}{9}\cdot\frac{7}{11}=\frac{28}{99}$

3. $\frac{3}{7}\cdot\frac{1}{15}=\frac{1}{35}$

4. $\frac{1}{12}\cdot\frac{1}{9}=\frac{1}{108}$

5. $\frac{7}{8}\cdot\frac{6}{7}=\frac{3}{4}$

6. $\frac{11}{20}\cdot\frac{11}{40}=\frac{121}{800}$

7. $\frac{2}{5}\cdot\frac{1}{5}=\frac{2}{25}$

8. $\frac{2}{5}\cdot\frac{5}{16}=\frac{1}{8}$

9. $\frac{3}{4}\cdot\frac{14}{15}=\frac{7}{10}$

10. $\frac{7}{8}\cdot\frac{7}{16}=\frac{49}{128}$

11. $\frac{1}{2}\cdot\frac{3}{16}=\frac{3}{32}$

12. $\frac{9}{20}\cdot\frac{1}{18}=\frac{1}{40}$

13. $\frac{1}{b}\cdot\frac{5}{a}=\frac{5}{ba}$

14. $\frac{1}{xy}\cdot\frac{7}{2w}=\frac{7}{2xyw}$

15. $\frac{3t}{7}\cdot\frac{5t}{14}=\frac{15t^2}{98}$

16. $\frac{t}{p}\cdot\frac{y}{x}=\frac{ty}{px}$

17. $\frac{ab}{cd}\cdot\frac{ef}{gh}=\frac{abef}{cdgh}$

18. $\frac{1}{f}\cdot\frac{h}{4f}=\frac{h}{4f^2}$

19. $2\frac{1}{7}\cdot 9\frac{5}{6}=21\frac{1}{14}$

20. $4\cdot 1\frac{7}{12}=6\frac{1}{3}$

21. $7\frac{3}{10}\cdot 7\frac{1}{16}=51\frac{89}{160}$

22. $7\frac{1}{7}\cdot 1\frac{2}{9}=8\frac{46}{63}$

23. $5\frac{4}{9}\cdot 8\frac{3}{7}=45\frac{8}{9}$

24. $9\frac{1}{9}\cdot 6\frac{4}{41}=55\frac{5}{9}$

25. $8\frac{1}{2}\cdot 2\frac{2}{15}=18\frac{2}{15}$

26. $9\frac{8}{9}\cdot 6=59\frac{1}{3}$

27. $10\frac{1}{12}\cdot 7\frac{3}{4}=78\frac{7}{48}$

28. $\frac{2}{3}$ of 500 = $333\frac{1}{3}$

29. $\frac{7}{8}$ of 2121 = $1855\frac{7}{8}$

30. $\frac{1}{22}$ of 44000 = 2000

31. $\frac{4}{7}$ of $5\frac{1}{3} = 3\frac{1}{21}$

32. $2\frac{5}{6}$ of 1100 = $3116\frac{2}{3}$

33. $1\frac{7}{8}$ of $4\frac{9}{13} = 8\frac{83}{104}$

Answers for 1-4 Assignments

1. $\frac{1}{9} \div \frac{2}{3} = \frac{1}{6}$

2. $\frac{4}{21} \div \frac{2}{7} = \frac{2}{3}$

3. $\frac{3}{7} \div \frac{1}{15} = \frac{45}{7} = 6\frac{3}{7}$

4. $\frac{1}{12} \div \frac{1}{8} = \frac{2}{3}$

5. $\frac{7}{8} \div \frac{6}{7} = \frac{49}{48} = 1\frac{1}{48}$

6. $\frac{8}{20} \div \frac{11}{30} = \frac{12}{11}$

7. $\frac{2}{5} \div \frac{1}{5} = 2$

8. $\frac{2}{5} \div \frac{15}{16} = \frac{32}{75}$

9. $\frac{3}{13} \div \frac{14}{15} = \frac{45}{182}$

10. $\frac{7}{8} \div \frac{7}{16} = 2$

11. $\frac{1}{2} \div \frac{7}{16} = \frac{8}{7} = 1\frac{1}{7}$

12. $\frac{9}{20} \div \frac{1}{12} = \frac{27}{5} = 5\frac{2}{5}$

13. $2\frac{1}{7} \div 4\frac{1}{6} = \frac{18}{35}$

14. $8 \div 1\frac{7}{12} = 5\frac{1}{19}$

15. $14\frac{3}{10} \div 7\frac{1}{16} = 2\frac{14}{565}$

16. $7\frac{4}{7} \div 5\frac{8}{9} = 1\frac{2}{7}$

17. $8\frac{4}{9} \div 8\frac{3}{7} = 1\frac{1}{531}$

18. $19\frac{7}{10} \div 6\frac{3}{5} = 2\frac{65}{66}$

19. $8\frac{1}{2} \div 20\frac{5}{14} = \frac{119}{285}$

20. $9\frac{2}{9} \div 6 = 1\frac{29}{54}$

21. $2\frac{1}{16} \div 7\frac{3}{4} = \frac{33}{124}$

Answers for 1-5 Assignments

1. 361

2. 2187

3. 1728

4. 256

5. 244.14

6. 164025

7. 30.19

8. 25

9. – 343

10. 6

11. 2

12. 3

13. 8

14. 15

15. 14.49

16. 100.22

17. .06

18. 6

19. 13.28

20. 4.31

21. 4.96

22. 3.88

23. 2.01

24. 1.85

Answers for 1-6 Assignments

1. 13

2. –8

3. 15

4. 45

5. 5

6. 49

7. 9

8. 39

9. 0

10. 9

11. 4

12. $\frac{17}{2}$

Answers for 2-1 Assignments

1. All real numbers
2. All real numbers
3. All real numbers
4. All real numbers
5. All real numbers greater than or equal to 0, or $w \geq 0$
6. All real number greater than or equal to –8, or $p \geq -8$
7. All real number except 7, or $x \neq 7$
8. All real number except $-\frac{8}{3}$, or $x \neq -\frac{8}{3}$
9. All real number
10. All real number except 11, or $x \neq 11$

Answers for 2-2 Assignments

1. $24x$
2. $5y$
3. $241g$
4. $22f + 40h + 2$
5. $15.73k$
6. $13x^2 - 23x + 5$
7. $24d^2 + 23c^2 + 2d + 3c$
8. $-2a^3 + 16ab + 19a - 11b$
9. $7x$
10. $7k + 70$
11. $-3k + 8$
12. $75h - 100$
13. $-r^2 - 22r$
14. $36x + 54w + 72y$
15. $-14x + 2.6y$
16. $6y^3 + 12y^2 - 72y$
17. $\frac{5}{21}h + \frac{2}{15}$
18. $6x^2 - x$
19. $-5x + 84$
20. $-2tk + 40th + 13tp$
21. $31m - 29n - 56$
22. 0

Answers for 2-3 Assignments

1. $x = 12$
2. $y = 4$
3. $t = -5$
4. $t = -\frac{111}{5}$
5. $x = \frac{97}{23}$
6. $t = 3.4$
7. $m = 86$
8. $y = \frac{60}{7}$
9. $x = 4.\overline{8}$

10. $w = \frac{55}{7}$

11. $r = 94.\overline{1}$

12. $d = -\frac{3}{32}$

13. $g = -\frac{1}{3}$

14. $r = -\frac{4}{15}$

15. $p = 5$

Answers for 2-4 Assignments

1. $w = 50$

2. $q = \frac{20}{3}$

3. $y = \frac{23}{3}$

4. $f = 168$

5. $x = 2.82$ or $\frac{138}{49}$

6. $h = 2.55$ or $\frac{130}{51}$

7. $m = -\frac{7}{43}$

8. $x = \frac{19}{8}$

9. $r = \frac{900}{97}$

10. $d = 119$

11. $k = \frac{137}{21}$

12. $h = -435$

13. $\frac{x\,quart}{56\,Liters} = \frac{1\,quart}{.95\,Liters}$, 58.95 quarts

14. $\frac{x\,ounce}{456\,grams} = \frac{1\,ounce}{28.35\,grams}$, 16.08 ounces

15. $\frac{x\,kilometers}{4.5\,miles} = \frac{1.61\,kilometers}{1\,mile}$, 7.25 kilometers

16. $\frac{x\,cubic\,meters}{460\,cubic\,feet} = \frac{1\,cubic\,meters}{35.32\,cubic\,feet}$, 13.02 cubic meters

17. $\frac{x\,cm^2}{2.08\,in^2} = \frac{6.45\,cm^2}{1\,in^2}$, 13.42 cm^2

18. $\frac{x\,gallons}{910.5\,pints} = \frac{1\,gallon}{8\,pints}$, 113.81 gallons

19. $\frac{x\,hours}{25.5\,seconds} = \frac{1\,hours}{3600\,seconds}$, .007 hour

20. $\frac{x\,tons}{9560.4\,kilograms} = \frac{1\,ton}{907.03\,kilograms}$ 10.54 tons

21. $\frac{x\,yd}{75.4\,meters} = \frac{1.09\,yd}{1\,meter}$, 82.19 yards. 82.19 yards ($36.50/15 yds.) = $200.00

22. a) $\frac{x\,gallons}{39600\,cubic\,inches} = \frac{1\,gallon}{231\,cubic\,inches}$, so $x = 171.43$ gallons of pool water,

b) $\frac{y\ gallons\ of\ pool\ cleaner}{171.43\ gallons\ \text{of pool water}} = \frac{1\ gallon\ \text{of pool cleaner}}{9\ gallons\ \text{of pool water}}$, so $y = 19.05$ gallons of pool cleaner, and $\left(\frac{\$2.00}{1\ gallon\ \text{of pool cleaner}}\right)(19.05\ gallons\ \text{of pool cleaner}) = \$38.10$

Answers for 2-5 Assignments

1. $r = 1764$
2. $w = \mathbf{162}$
3. $c = \frac{10}{11}$
4. $k = \mathbf{27}$
5. $p = \frac{17}{9}$
6. $t = \mathbf{660}$
7. $n = -\frac{651}{50}$
8. $h = -\frac{17}{2}$
9. $p = \frac{225}{14}$
10. $g = \frac{22}{75}$
11. $a = \mathbf{2}$
12. $b = \mathbf{-72}$
13. $x = \frac{380}{7}$

Answers for 2-6 Assignments

1. **a)** $y = \frac{5}{12}$ **b)** $p = 132$
2. **a)** $r \approx 9.57$ **b)** $y = -5.27$
3. **a)** $x = \frac{19}{2}$ **b)** $y = -\frac{29}{12}$
4. **a)** $q = 7$ **b)** $p = 5$
5. **a)** $y = -1.1$ **b)** $A = 0$
6. **a)** $r \approx .085$ **b)** $p \approx 441.67$
7. **a)** $h \approx 19.89$ **b)** $A \approx 178.98$
8. **a)** $h = -221$ **b)** $k = 33$
9. **a)** $w = -9$ **b)** $y = 4$ **c)** $x = 3.5$ or $= \frac{7}{2}$
10. **a)** $V \approx 7.07$ **b)** $p = 1033.75$

Answers for 2-7 Assignments

1. $x = \frac{y-2}{3}$
2. $y = \frac{1-r+x}{2}$
3. $b = \frac{A}{.5h} = \frac{2A}{h}$
4. $y = \frac{n-12t}{2r}$
5. $p = 8t + 2K + 2$
6. $x = 2t + h$

7. $w = \dfrac{(24y - x)}{32}$

8. $h = \dfrac{V}{3.14r^2}$

9. $t = \dfrac{2x - kh}{2h}$ or $= \dfrac{x}{h} - \dfrac{k}{2}$

10. $x = 3a + 3b - y$

11. $m = \dfrac{3d + 7b - 9k - 14g}{7}$

12. $c = \dfrac{a + 2b}{48}$

Answers for 2-8 Assignments

1. .625 = 62.5%
2. 12%
3. 5600
4. 200%
5. 840636.9
6. 5.1%
7. 66 games
8. 2.345
9. 7.5%
10. 5,200 lightbulbs
11. $2.03
12. $5,785
13. $29,750.00
14. $183.33
15. 75%
16. a) 23.5% b) 235 customers
17. 69%
18. $1815.79
19. 71.4%
20. 400 units

Answers for 3-1 Assignments

1. True
2. False
3. True
4. True
5. False
6. True
7. True
8. False
9. Inch, centimeter
10. Foot, meter
11. Fluid ounce, millileter
12. Pound, kilogram
13. Quarts, liters
14. Gallon, liter
15. Pound, kilogram
16. Inch, millimeter

17. Ounce, milligram
18. Foot, centimeter
19. Ounces, grams
20. Tons, megagram
21. Centi
22. Kilo
23. Milli
24. Deca
25. Mega
26. Micro or μ
27. Hecto
28. Deci

Answers for 3-2 Assignments

1. 12.25
2. 147
3. 11880
4. 2.25
5. 3.094
6. 396
7. 218
8. 3.375
9. 22 ft 1 in
10. 9 mi 1290 feet
11. 18 gal
12. 9 gal 3 qt
13. 10 hr 29 min
14. 24 min 50 sec
15. $\frac{261 \text{ mile}}{1 \text{ hour}}$
16. $\frac{70 \text{ lbs}}{1 \text{ in}^2}$
17. $\frac{48 \text{ mi}}{1 \text{ gal}}$
18. $\frac{946 \text{ yds}}{1 \text{ hour}}$
19. $7.3\overline{3}$
20. $10.6\overline{6}$
21. 2925
22. 11.6875
23. 320
24. 22.56

Answers for 3-3 Assignments

1. 16
2. 176
3. 32000
4. 13
5. 9.75
6. 21120
7. 4048
8. 41
9. 4.5
10. 512
11. 1712
12. 1296
13. 7
14. 559872
15. 149.61
16. 10.80
17. 506880
18. 419904
19. 201.97
20. 29.92
21. 1.80
22. .56
23. 57.5
24. 33123600
25. 758.33
26. .0056
27. 8910
28. 20.53
29. 11781.82
30. 30.67

Answers for 3-4 Assignments

1. .5
2. .023
3. .624
4. 45
5. 4350000
6. 307000
7. .065
8. 800
9. 43.4
10. .05
11. .00596
12. 5000
13. 3310
14. .009
15. .45
16. 23.456
17. .0123
18. 920
19. 123000000000
20. 345

Answers for 3-5 Assignments

1. .0001
2. .000001
3. 10000
4. 1000000
5. 150000
6. 3.42
7. 56.7
8. 11000000
9. 16.23
10. 109.75
11. 1.83
12. 45.42
13. 32.18
14. 1860.33
15. 6001.65
16. 809.24
17. 3.22
18. 1132131.76
19. 75700
20. 1.71
21. 484.43
22. 2.64
23. 19.45
24. 1.07
25. 17003.96
26. .11
27. 10543.59
28. 48.82
29. .59
30. 3.52
31. 1.63
32. .08

Answers for 3-6 Assignments

1. 20 milliliters of the 9% solution and 30 milliliters of the 19% solution.
2. .4 quart of regular milk and .6 quart of chocolates milk.
3. 24 ounces of "Mildly Garlic" and 16 ounces of "That's so Garlic".
4. 7.5 fluid ounces of solution A and 7.5 fluid ounces of Solution B.
5. 1.8 pounds of red jellybeans and 1.2 pound of green jellybeans.
6. 19.2 fluid ounces of salad dressing A and 12.8 fluid ounces of salad dressing B.
7. 4 FancyNail sets.
8. 1 pound of Cashews and 3 pounds of Peanuts.

Answers for 4-1 Assignments

5:40 PM, 1850, 6:35 PM

12:40 AM, 2311, 8:30 PM

7:20 PM, 2210, 1:05 PM

1755, 3:10 PM, 0815

11:10 PM 1:15 AM, 1555

1905, 3:40 AM, 5:45 PM

7:46 PM, 6:19 PM, 1355

1559, 12:09 AM, 2020

2:35 AM, 1433, 6:55 PM

5:05 AM 1040, 12:00 AM

0111, 1311, 12:00 PM

Answers for 4-2 Assignments

1. 5 hours 57 minutes
2. 7 hours 48 minutes
3. 7 hours 15 minutes
4. 43 minutes
5. 1 hour 13 minutes
6. 12 hours 57 minutes
7. 2 hours 56 minutes
8. 4 hours 23 minutes
9. 22 hours 57 minutes
10. 7 hours 59 minutes
11. AM 3 hours 51 minute PM 3 hours 46 minutes Total 7 hours 37 minutes
12. AM 2 hours 50 minutes PM 2 hours 49 minutes total 5 hours 39 minutes
13. AM 4 hours 4 minutes PM 4 hours total 8 hours 4 minutes
14. AM 4 hours 30 minutes PM 4 hours 15 minutes total 8 hours 45 minutes
15. AM 3 hours 47 minutes PM 2 hours 37 minutes total 6 hours 24 minutes
16. 12:30 PM
17. 9 hours 52 minutes

18. 8 hours 43 minutes

19. 10 hours 20 minutes

20. 1:30 PM

21. 11:40 AM: 5:10 PM: 8 hours 40 minutes

Answers for 4-3 Assignments

1. 15 hours 13 minutes
2. 12 hours 7 minutes
3. 25 hours 24 minutes
4. 21 hours 9 minutes
5. 146 hours 47 minutes
6. 8 hours 22 minutes
7. 9 hours 35 minutes
8. 23 hours
9. 12 hours 29 minutes
10. 7 hours 10 minutes
11. 19 hours 22 minutes
12. 21 hours 26 minutes
13. 3 hours 10 minutes
14. 5 hours 35 minutes
15. Yes, the total is 6 hours 40 minutes so fits in 8 hours, time left is 1 hour 20 minutes
16. The total is 4 hours 5 minutes, so there is 55 minutes for break time.

Answers for 4-4 Assignments

1. $6\frac{11}{20}$ hr
2. $1\frac{1}{60}\,hr$
3. $6\frac{59}{60}\,hr$

4. $2\frac{1}{6}hr$

5. $19\frac{2}{5}hr$

6. $5\frac{3}{4}hr$

7. $2\frac{1}{60}hr$

8. $11\frac{4}{15}hr$

9. 60 hr

10. $1\frac{1}{20}$ hr

11. 7 hr 25 min

12. 1 hr 1 min

13. 35 min

14. 9 hr 52 $\frac{1}{2}min$

15. 112 hr 11 $\frac{1}{4}min$

16. 160 min

17. 742 $\frac{1}{2}min$

18. 295 min

19. 13252 min

20. 457 min

21. $9\frac{1}{2}hr; 9\,hr\ 30\,min$

22. $14\frac{11}{30}hr; 14\,hr\ 22\,min$

23. $17\frac{1}{6}hr$; 17 hr 10 min

24. $12\frac{8}{15}hr; 12\,hr\ 32\,min$

25. $22\frac{11}{60}hr; 22\,hr\ 11\,min$

26. $13\frac{71}{120}hr; 13\,hr\ 35\frac{1}{2}min$

27. A) $8\frac{47}{60}hr; 8\,hr\ 47\,min$

B) $1\frac{17}{60}hr; 1hour\ 17min$

28. July 6: $6\frac{21}{40}hr; 6\,hr\,31\frac{1}{2}min$

July 7: $5\frac{1}{4}hr; 5\,hr\ 15\,min$

July 8: $3\frac{39}{40}hr; 3\,hr\,58\frac{1}{2}min$

July 9: $6\frac{3}{10}hr; 6\,hr\,18min$

July 10: $7\frac{9}{40}hr; 7hr\ 13\frac{1}{2}min$

Total: $29\frac{11}{40}hr; 29\,hr\,16\frac{1}{2}min$

Answers for 5-1 Assignments

1. $35.18; $2.18
2. $24.27; $2.27
3. $23.65; $2.15
4. $946.00; $66.00
5. $10.40; $0.50
6. 8%
7. $12.00; $13.08
8. 10%
9. $0.31
10. $350.00
11. 11.5%
12. $18.23
13. $37.06
14. $1085.91
15. 11.0%

Answers for 5-2 Assignments

1. $8.53; $24.03
2. $22.60
3. $10.50
4. $20.35
5. $58.90; $20.90
6. $3.63
7. $18.56
8. $35.31
9. $37.13
10. $15.15
11. $37.54
12. 120%
13. $34.78
14. $16.80; $27.72
15. $14.60; $13.87

Answers for 5-3 Assignments

1. $8.53
2. $20.00
3. $24.84
4. $1250.00
5. $15.34
6. 25%
7. $672.00
8. $315.00
9. $7.48
10. $14.21
11. $10.03
12. $18.17
13. $7.50
14. $17.70
15. $9.22
16. $18.94
17. $24.50

Answers for 5-4 Assignments

1. $3260.42
2. $42000.32
3. $13050
4. $29000

5. $595
6. $22.56 a bottle, total sales would be $2707.20
7. $29.00 a jar, total profit $870.00
8. 5 massages
9. $60 original price, $42 discounted price
10. $1565

Answers for 5-5 Assignments

1.

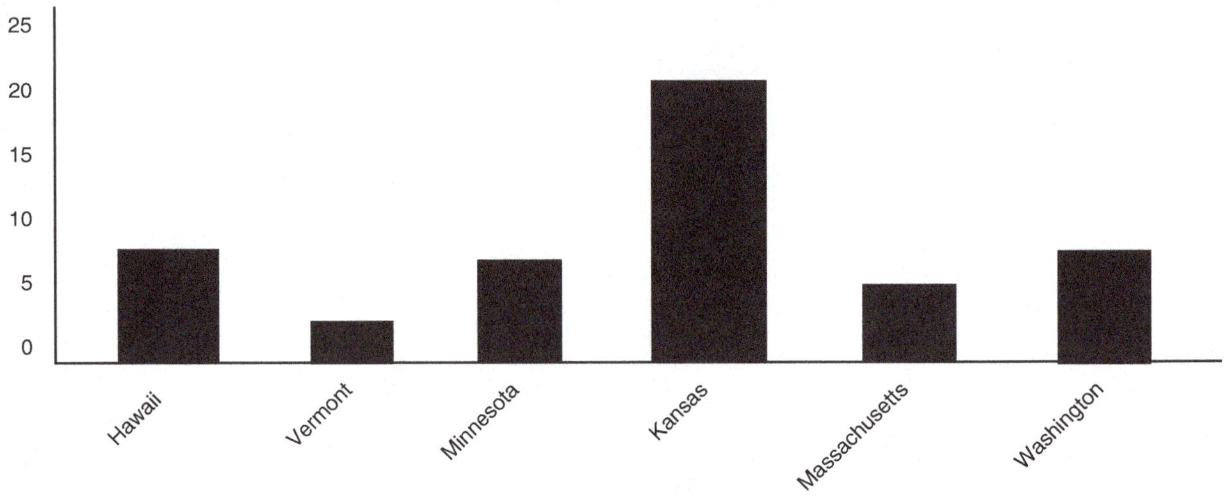

2. (20% is 72°, 30% *is* 108°, 50% *is* 180°)

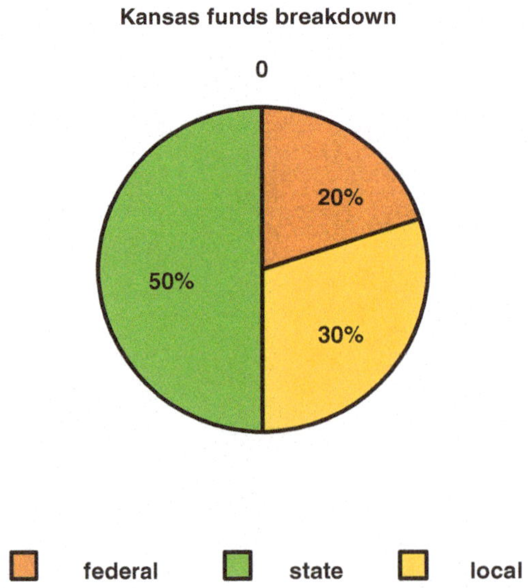

3. Frequency table

Interval in sec.	0 - 2.5	2.5-5	5-7.5	7.5-10	10-12.5	12.5-15	15 - 17.5	17.5-20
Frequency	1	8	12	6	5	3	0	1

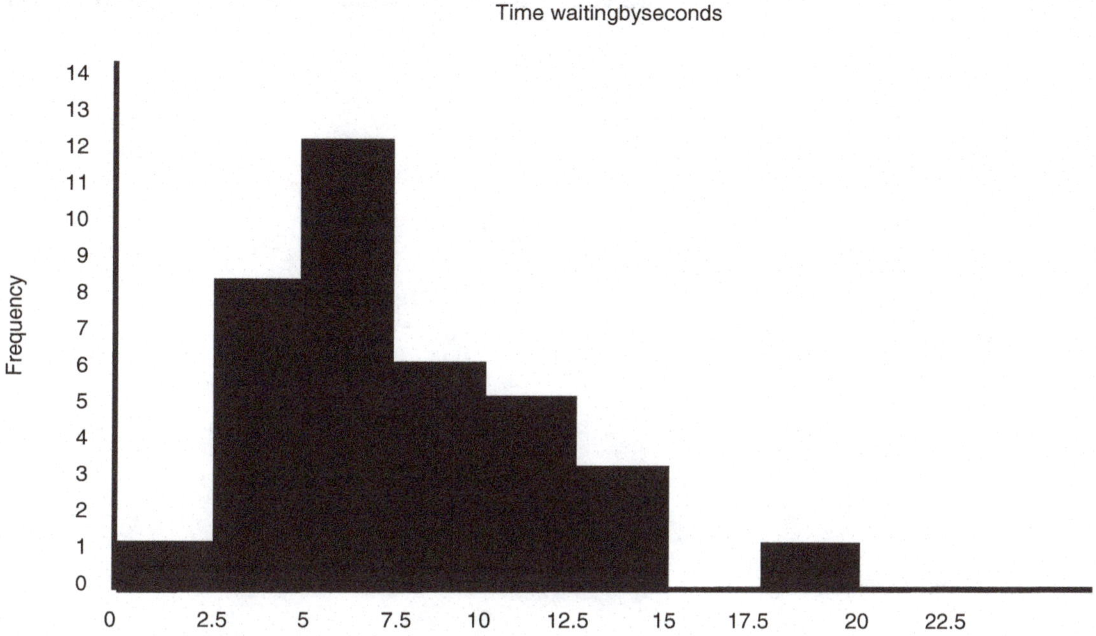

4. Hair product, Number of votes, % of vote : Aveno 30, 20% ; L'Oreal Paris 45, 30% ; Garnier 42, 28% ; Dove 18, 12% ; Clairol 15, 10%

5.

6. **a)** 4

b) 0

c) 13

d) 20

7.

hours	0	1	2	3	4	5	6	7	8
cost $	$120	$170	$220	$270	$320	$370	$420	$470	$520

hours	0	1	2	3	4	5	6	7	8
Sales $	0	$70	$140	$210	$280	$350	$420	$490	$560

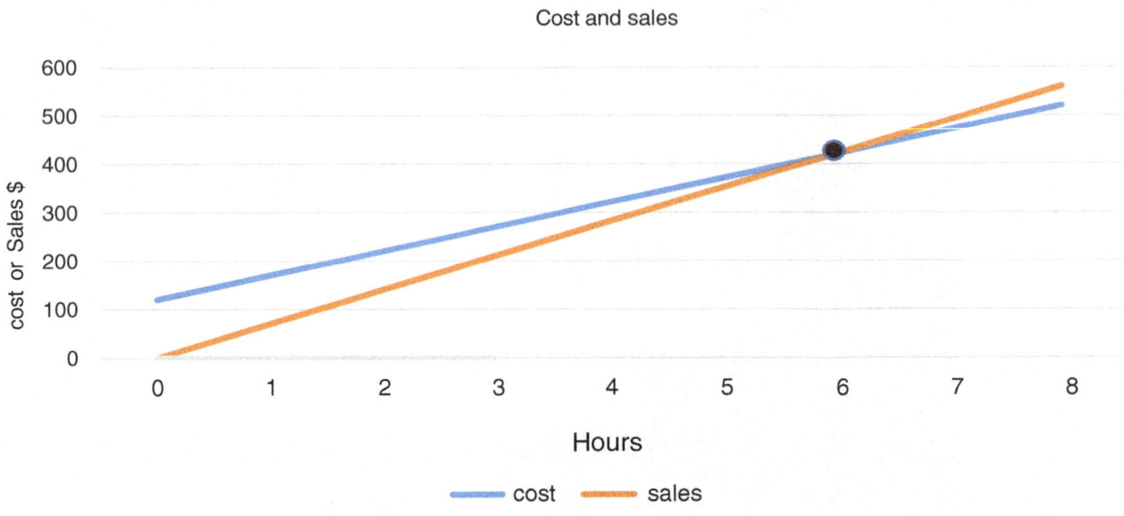

Breakeven point is 6 hours, $420 in cost or sales

8. 2019 Total Sales $175,000, Total Fixed costs $40,000, Total Variable cost $105,000

Average Sales per client is $50.00

Average Variable cost per client is $30.00

Sales and costs Tables

clients	0	500	1000	1500	2000	2500	3000	3500	4000	4500	5000
Sales $	0	$25,000	$50,000	$75,000	$100,000	$125,000	$150,000	$175,000	$200,000	$225,000	$250,000

clients	$	500	1000	1500	2000	2500	3000	3500	4000	4500	5000
cost $	$40,000	$55,000	$70,000	$85,000	$100,000	$115,000	$130,000	$145,000	$160,000	$175,000	$190,000